# YOU CAN TEACH
# ABOUT NUMBERS

ALSO IN THIS SERIES:

*Bad Behaviour, Tantrums and Tempers*
*Families and Friends*
*Fighting, Teasing and Bullying*
*Food: Too Faddy? Too Fat?*
*Good Habits, Bad Habits*
*Growth and Development*
*Stammering in Young Children*
*Successful Potty Training*
*Teach Your Baby to Sleep Through the Night*
*Worries and Fears*
*You Can Teach Your Child to Read*

BY THE SAME AUTHOR:

*You Can Teach Your Child to Read*

Adrienne Katz

# YOU CAN TEACH YOUR CHILD ABOUT NUMBERS

## Learning is a Game We Play with our Minds

Thorsons
*An Imprint of* HarperCollins*Publishers*

Thorsons
An Imprint of HarperCollins*Publishers*
77–85 Fulham Palace Road,
Hammersmith, London W6 8JB
1160 Battery Street,
San Francisco, California 94111-1213

Published by Thorsons 1994
1 3 5 7 9 10 8 6 4 2



Illustrations by G.M. Motley



A catalogue record for this book
is available from the British Library

ISBN 0 7225 2849 3

Printed in Great Britain by
HarperCollinsManufacturing Glasgow

For Liane and Ian, who outstripped me and soared ahead so quickly.

# Contents

*Acknowledgements* ix
*Note* x
*Introduction* xi

**1 A Word to Parents** 1
Questions for us 1
Learning is making sense of things 3

**2 Laying a Foundation: Before Numeracy** 7
Matching, sorting, pairing, contrasting 7
Shapes 10
Patterns and symmetry 12
Language for learning and thinking 16

**3 Making Friends with Numbers** 21
Experience before symbols 21
Counting 26
Getting to know the shapes of numbers 31
Introducing counting games 33
Using numbers in real life 37

**4 Adding and Subtracting** 40
Number lines 42
Hundreds, tens and units 46

**5 Multiplying and Dividing** 50
Multiplication tables 53
Next steps—real life 58

**6 Measuring** 60
Length 60
Perimeter and area 63
Scale 64
Capacity and volume 66
Weight 69
Time 72
Direction 75
Angles 78
Recording measurement 82

**7 Fractions, Decimals and Percentages** 84
Fractions and decimals 84
Percentages 87

**8 More Games, Puzzles and Challenges** 90
Be observant 90
Adding and subtracting games (ages 5 and up) 91
Sequence 95
Pattern puzzles (ages 6 and up) 95
Symmetry 97
Origami 99
Tessellations 103
Carmaths (ages 6 and up) 108

*Appendix 1* *Counting Rhymes* 112
*Appendix 2* *Hands-on Fun* 117
*Appendix 3* *Books to Help with Counting* 122

*Index* 125

# Acknowledgements

• • • • •

My thanks to the children and their teachers who shared their experiences with me.

# Note

• • • • •

It is an age-old problem whether to use he or she when talking about a child. To repeat both each time is clumsy. Always to refer to 'him' is unfair and irritating to the girls. I hope you will be understanding about the changing gender of the child from chapter to chapter.

Age—Is an activity appropriate for your child?

You and I know that no two children are alike. What one five-year-old finds old hat another has yet to meet. So please be guided by your own child's readiness rather than by chronological age. If he does not enjoy one of the games in this book, leave it for another time.

# Introduction

• • • • •

I sat in the classroom among the five-year-olds fascinated by their reasoning, their curiosity and logic. But I left at the end of the day deeply worried. I had found that in an average first-year reception class there could be as much as 18 months' difference in development. This was the end of the second term. Where Sam was already able to subtract and work out how many pennies he needed to do some simple shopping, Joanne and Tommy still could not count and, more worrying still, they did not understand the *concept* of counting.

They were asked to arrange a tea party for the three bears, and shown into the playhouse corner complete with tea-sets, chairs and teddies. They had a wonderful time laying out every cup and saucer they could find, then all the plates and soon everything else available. When we had a little chat about the bears—how many of them were there? which cup was for Daddy bear? which for Mummy bear and which for Baby bear? and several other careful questions—their lack of understanding became even more apparent. They had no idea of ordering—Daddy bear, Mummy bear and Baby bear, for instance, or of one-to-one correspondence—a cup for each bear.

If your child comes into school without the slightest idea of numbers, and their use in our daily lives, he will have a great deal of work to do to catch up with others, who are of course racing ahead, making the gap ever wider. The slower child sees himself as stupid within a short time and may begin to believe that he cannot do what is asked of him.

A child like Sam will delight in his success and his teacher's approval. Joanne, on the other hand, might begin to look out of the window, or keep quiet rather than try when it all seems too confusing. Her teacher may become discouraged by her lack of interest and begin to expect less from her. Because of her lack of readiness, she will be slow to learn and a cycle of being unable to keep up may set in. Because she fails she will lose heart and stop trying. Sam and Joanne will have very different educational experiences, yet they probably started out with similar innate ability.

The busy teacher with a group of 20 and often more, will be hard pressed to find the time to spend bringing your child to the starting point. Those who have had preschool education and a great deal of parental input might be more than a year ahead, and will have the tools they need in terms of language and understanding to explore the new ideas the teacher will introduce.

So what does a child need? In my book *You Can Teach Your Child to Read*, I have discussed a child's needs for language and reading. Here we are thinking about maths. A child needs to be friends with numbers, recognize them, know their names and understand their use in everyday life. Above all he needs to understand the first two stages of the concept of counting (see pages 26-29).

He also needs the tool of language. He needs the words to be able to think and ask questions. Maths is not only about doing sums. It is about patterns, logic and ideas. To investigate, explore and discover we need words. If we ask, 'What do you notice about that?', a child might need to classify or compare to answer. He needs the words that describe what he finds, 'This one is longer than the red one,' 'There are more sweets in this packet.'

But, you may be thinking, I am not qualified to teach, how will I know what he needs or how to do this? Be reassured. All your child needs you can provide. It is talking together as you go about your daily routine...playing together and singing those number rhymes from your childhood, counting as you lay the table or go shopping. Weaving numbers into the fabric of your life together is easy because their use is so integral, remembering to talk and show is a habit most parents of young children soon acquire.

On the other hand, you may be one of those parents who reads this and thinks—this is surely stating the obvious, I need more than this. 'Read further' is my plea. This book can be used in depth, with its many suggestions for creative games and activities you might build upon or develop. It can also be taken as a simple guideline to a way of doing things with your children that will enhance their pleasure in the world of numbers.

For parents of those children like Joanne and Tommy whose lives seem to lack these parent/child conversations and early explorations—I hope I can convince you that it is not too late, and truly vital to your child's future that some adult spends one-to-one time with him, helping him to discover these interesting things called numbers and, through talking, building up his vocabulary and understanding in a way that TV cannot do. Parent power is a very precious commodity. To you, your child is unique and you can explain when he asks a question, explore when he is curious and of course stop when he seems bored or restless. You work to no timetable and can pick the right moment because you know your child and are guided by him.

The danger in a book of this type is that an overeager parent could use it to push a child to perform like a hothouse plant. Could I be forgiven, therefore, for setting out some guidelines here?

Don't be a taskmaster, don't make finding out about numbers a chore, sternly insisted upon. Don't stop a child who is playing to 'come and do some number work.' Avoid comparing your child with any other. Make no demands for repayment, ask nothing in return. Too many parents invest so much in 'teaching' their child that they are looking for some reward, i.e. a brilliant school report, or reflected glory. Some of us are doing this subconsciously.

To teach is to give a gift with no obligation.

# 1

. . . . .

# A Word to Parents

## QUESTIONS FOR US

What do kids need from maths? How do we answer questions such as 'Why do I have to learn this?', and how does a parent manage who has no confidence in her own maths?

These are questions circling around me as I write. There are a couple more, like 'How do I find the time?' and 'How do I make my child concentrate?' that have to be acknowledged.

We are after all only human. Most parents are stretched to the limit a great part of the time. If only we could have long days stretching ahead without other commitments, we could see ourselves doing all these creative things with our children. This is said to me so many times a month by parents I meet as I work or visit schools. Well, I know the pressures, I too am a working mum. But some of the rushing around and treadmill schedules are not truly necessary. Spending time with our children is. It may mean a rethink and a change of priorities, but time has to be carved out or they will have grown and left home before you have spent some intimate time together.

On the question of our 'mathsphobic' generation, it is true that a great number of parents fear maths, or feel they are simply no good at it. In the National Child Development Study, 39 per cent of the people reporting literacy or numeracy problems reported difficulties with numeracy. The idea is not to pass this on to another generation. Confidence is vital. Familiarity, meeting these concepts through play in a relaxed environment like home, may help prepare our kids for a future in which understanding mathematical structure and being competent at calculating serve them well.

If children can see us using mathematical ideas they will know these ideas are useful for everyday life. They will gradually grasp the use of maths for science, industry and getting systems such as trains, buses or airplanes to work. With our help they may realize that maths allows us to explain, describe and predict...they will develop logical thinking and some will enjoy its aesthetic appeal, finding a more elegant method to solve a problem, or noticing a pattern. Mastering this could be satisfying. Above all, maths could lose its mystique and, because they have fun with it, they won't fear it.

Your ability is not a problem. You can explore together and find out. Your role is not to be the teacher—your child has one of those at school. You are a fellow traveller who makes discovery possible. A sort of enabler, if you like, who creates learning opportunities from everyday situations, who will play a game for the fun of it, share the wonder and curiosity of a new mind meeting all this for the first time. You may come to enjoy maths yourself.

You may also learn from your child as she streaks ahead leaving you far behind, as mine have done, laughing gently at the very idea of Mum writing about maths. There is nothing wrong in this. A parent's role is to nurture a child until she can cope in the adult world, not necessarily to follow where she goes every step of the way.

So, if we help with the very beginnings of her maths learning, well within our grasp, does it matter if she soars higher where we cannot follow? Our children will study many subjects that we don't know a thing about; we have only to give them a good start. Let me share a small secret with you. I know children feel marvellous when they discover that they have outstripped their parents at something. It does their confidence the world of good to know this. That is why when the moment comes you can praise and encourage them, knowing that you laid the foundations. You may have to put up with a little scoffing and jeering at your own performance. But I do this happily. (I am also let off difficult homework, for which I've been grateful for years.)

For those who say they can't get their child to settle or concentrate—don't make it a battle. Leave the idea of sitting still altogether. Take other opportunities as they arise in your daily routine to talk and play. If your child is resistant or reluctant, leave it alone for a few weeks. Follow her lead and look at what interests her at the moment. A 22-month-old visitor, Alexander, showed me very firmly yesterday that he wasn't interested in stacking plastic bowls inside one another, or finding their lids. He wanted to study how the step of the kitchen stool folded. This he did over and over again. Then he tipped the stool onto the floor and tried folding up the step from another angle. He kept looking at me for my reaction. All he needed was for me to make this exploration possible. I smiled and said a few words like 'Open, closed' or 'That's clever!' I stayed nearby for company and watched him, making eye contact each time he looked up to check. He would have been rightly angry had I whisked him off the floor to 'Come and do some mathematical play.' He was observing, testing and finding out like any scientist.

## LEARNING IS MAKING SENSE OF THINGS

Watch that toddler. She drops the plastic ring and is fascinated by the way it wobbles and then lies flat. She lifts it up and drops it again, repeating this experiment for herself over and over again. She is creating knowledge out of what she knows and experiences.

As she observes and wonders she is testing and finding out. She is making sense of things. Adults sometimes seem to knock this curiosity out of children in their efforts to make them behave as they believe they should, in particular in their efforts to keep the home and their lives in it in some sort of order. This is not to suggest that you should live in chaos because you have a two year old. But we can create situations in which our children can experiment in safety. We can offer opportunities to children to satisfy their natural curiosity, rather than stifle it.

We can be guides on a voyage of discovery, share their delight in the most mundane sight. This may mean standing transfixed in the middle of roadworks many times because a child wants to watch the dump truck complete its manouevre for the umpteenth time.

It means allowing a child the time to watch—not to be always hurried away. It means offering the simple basic toys such as wooden bricks with which your child can experiment. By placing one brick on top of another or in a line, size, shape and weight begin to be apparent. Building towers or fitting shapes into slots, watching the ball roll away—all help learning through understanding in a way that some elaborate toys never allow. Children also need access to ordinary everyday household items. They see these as your toys, so will want to clash the saucepan lids or play with a wooden spoon as (to them) you do.

Above all, I think, being your child's guide means respecting her explorations. Play is a serious business and adults often interfere, knocking a game off course or diverting a child from her path. Overeager interference can make a child stop that activity altogether. If we respect a child's mind, we come to see that the amazing feat of learning that takes place in the early years is a phenomenal achievement. This deserves polite alertness and interest from us, but not swamping over-enthusiasm.

There is another aspect of adult behaviour that can be a

problem. We naturally praise our children and are thrilled with every step along the ladder of progress. Praise helps build self-esteem and is so valuable we should not waste it. However, if a child is praised inappropriately for everything she does, she may become dependent on our approval. She is not then learning to satisfy herself but to gain our approval. A very conscientious child can even come to fear the loss of this approval so much that she becomes afraid of doing the wrong thing and less inclined to experiment. There is a fine line here, between losing sight of the real goal or being too harsh and not praising enough. Yes, it's another of these difficult parenting situations: what is enough? Sensitivity is probably the most precious gift in this relationship.

Naturally you will know that not all children develop at the same pace. You'll know your own child and will somehow resist comparing her progress with that of every other two year old you know. Some walk and talk late, some early. All healthy children walk and talk in the end.

As your child learns from experience, she will ask questions. Some of these you will be able to answer. But if like me you have very inquisitive children, you will often not be able to explain how sound is held on cassettes or why the world is as it is. Please do not answer 'Because it just is' in frustration too often. We can still learn from experience ourselves, and finding out the answers to my children's questions taught me (and still does) more than years of school.

It simply means finding someone who can explain it, or looking it up somewhere. This creates a habit of searching for information which will help your child throughout life. Besides, it is fascinating as an adult to learn about holes in Gruyère cheese, satellites in orbit, birds' egg-laying habits and much, much more. I may have enjoyed discovering the world for myself as a child, but I enjoyed seeing it anew *with a child* far more.

A brief outline of early action:

- Look at things together, from great big glorious peonies to the gears in the music box. Talking and showing costs nothing.
- Allow your child some quiet time to contemplate and think. Don't overdo the talk at all times.
- Teach her big and small. Arrange bricks in order of size, or build a tower with the big ones at the base. Fit cups or saucepans one inside the other.
- Look at bridges in toy trains or real life. Can she build a bridge with a few wooden bricks?
- Sort buttons, acorns, pebbles, cutlery, groceries, coloured wooden bricks, animals, vehicles...
- Gradually introduce the idea of pattern.
- Compare and distinguish.
- Introduce the true value of low numbers: one, two and three.
- Use rhymes and songs, finger play and tickles (some fun counting rhymes are in Appendix 1).

# 2

. . . . .

# Laying a Foundation

## BEFORE NUMERACY

## MATCHING, SORTING, PAIRING, CONTRASTING

Before such an abstract idea as number can develop in your child's mind, he needs to have had broad experience of sorting, matching, distinguishing and ordering real things. Each step in understanding is built on the one before and, as the early years are so vital for the development of intelligence, unrestricted play and experiment allows your child to behave as a scientist does: observing, testing, comparing and describing.

As your child begins to develop vocabulary you will notice the way he classifies what he labels. One of my children was crazy about cars. At first everything with wheels was a car and 'Me go car' the sentence of the moment. Gradually, however, he noticed aircraft in the sky and in picture books. After much thought he told me they were 'birdycars'. Clearly they flew like birds but were in another sense like cars, transporting people. His father's passion for vintage cars he saw as Daddy playing with toy cars, and soon defined these as 'baby Daddy cars' which, of course, they were. Your child may first use 'car' for all vehicles and then come to distinguish between lorries, buses and taxis.

In the same way, many children call all birds 'duck' and all animals 'dog', but they gradually perceive subtle differences in ears, tails and sizes. Distinguishing differences is a mathematical skill. Two small boys I have known could tell me every make of car passing in a busy road in their third year. This is observing at its most refined. The child is noticing little differences in the design of cars. He is unfazed by colour, but looks instead at shapes. This is in effect the same skill as telling the difference between the letters of the alphabet or between numbers.

This is why any games which involve sorting, matching, comparing and looking for similarities and differences are valuable. Select and discard from a pile of objects to form a set of like items. Describe the objects. In matching we are looking for a sameness, a common property. This will lead to games of Snap and variations on it. Sort by colour, by size, by features, numbers (i.e. buttons with two holes in this pile, buttons with four holes in the other pile).

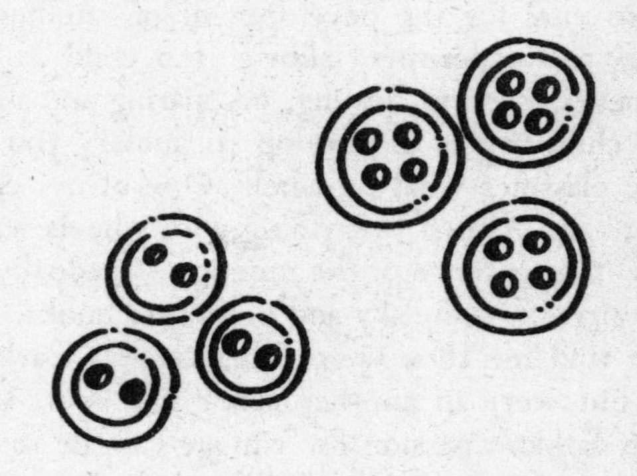

## Can You Find?

- From a pile of coins, hand your child one. Can he find another that is the same as it? Can he build a stack of others that are the same size?
- Look for soup cans, cartons, matchboxes and other things that can be sorted by size.
- Sort cutlery, sort laundry—socks into pairs, family members' piles—match cards, dominoes, sticks the same length, wooden beads the same shape, Lego bricks that are the same, lengths of wool, plastic letters of the alphabet.
- Contrast and look at differences. Talk about what you can see on a walk: buses, trucks and vans; different types of dogs; people with jeans and those without.
- If you have a big bucket of bricks, sort all the same-shaped bricks together. Try this with acorns, conkers, cones or pebbles—or use gummed paper shapes if you like. Each group of objects that have something in common are grouped in a set. Place these within a circle made of string or an old hula hoop. Interesting questions emerge automatically from this. For example, are there more acorns in the acorn set than conkers in the conker set? How many more?

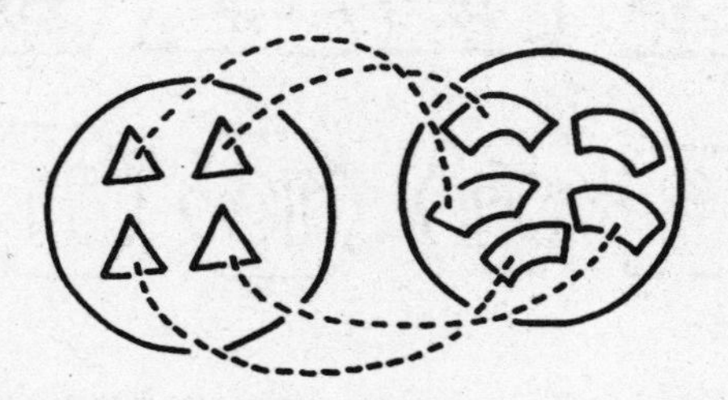

*There is one more strip than there are triangles*

## SHAPES

Look closely at shapes. The diagram below shows the mathematical names of several shapes. These can be found at home and outside. What shape are cereal boxes, or letter boxes? What shape is the cardboard core of the paper towel roll?

### Shapes by Touch

In this game, each player is blindfolded and guesses the answer by touch. If your child hates being blindfolded, put the objects into an old woolly sock or beneath a little blanket or flannel, or even in a special 'mystery shoe box' with a hole at one end big enough to fit his hand and retrieve objects from.

First choose things at random and have your child recognize everyday items, such as a comb, a coin, a toothbrush and a teaspoon.

Introduce shapes you have talked about, like squares and cylinders. Can your child recognize these unseen?

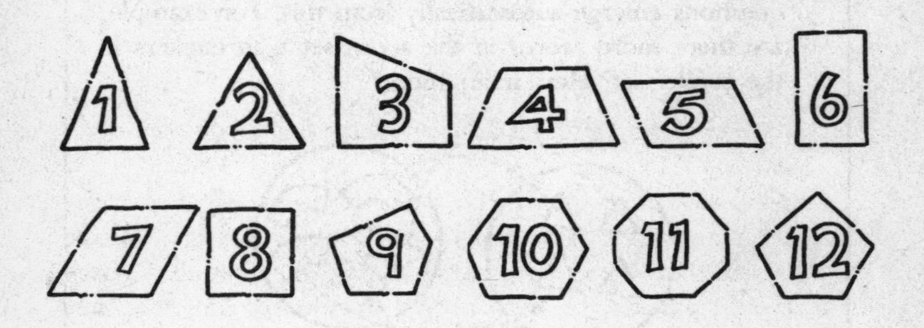


*1. isosceles triangle; 2. equilateral triangle; 3. quadrilateral; 4. trapezium; 5. parallelogram; 6. rectangle; 7. rhombus; 8. square; 9. pentagon; 10. hexagon; 11. octagon; 12. regular pentagon*

## Stranger

Find three items that are the same shape and one that is different, such as three marbles and one wooden brick. The object is to spot the one that is different. It does not form part of the set. The child feels in the sock and calls out 'The square is the stranger.'

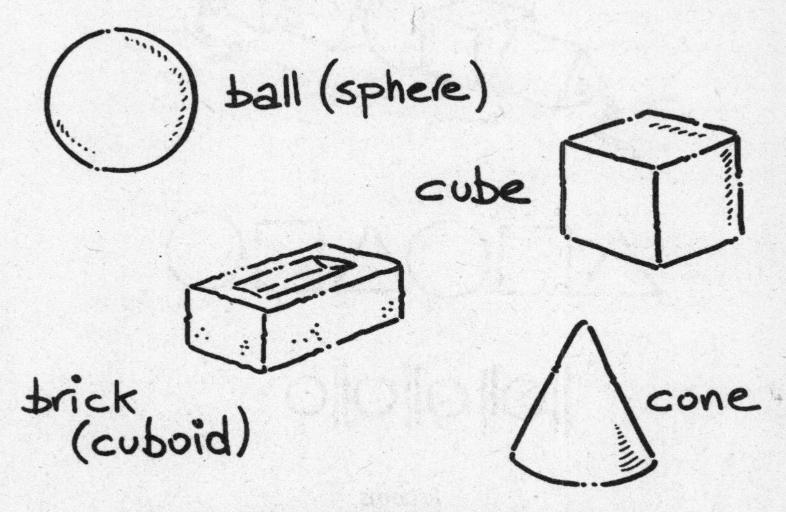


***Hunt for these shapes at home and when you are out***

At every stage new discoveries need the language to describe them. Living with a three year old makes your jaw ache, as the eternal question 'Why' is at its peak at this age. After all the baby experience, parents are hit by such a wave of thinking that they need to be on their toes with answers every waking hour.

Play all the variations you can with shapes. Name them, match them in pairs, sort and classify them. To do this you will need to find a name label for each set of objects.

## PATTERNS AND SYMMETRY

Why is pattern important in maths? What on earth have prints of bricks or rubbings of coins got to do with maths?

Pattern is important because it has to do with relationships. Relationships lie at the root of maths.

*Patterns*

At first children come to recognize pattern and to predict what comes next in a sequence. We thread up some coloured beads on a shoelace, the colours go red, blue, green, red, blue...? Invite a young child to continue and he will soon be threading away repeating this pattern once you have pointed it out and talked about it together.

If a number of beads are arranged in alternate colours—say red, yellow, red, yellow—he will be able to tell what colour the next bead should be if the pattern is to be followed.

## Patterns Without Numbers (Ages 3 to 7)

In this game, a fly on the wall would see you leaping about, waving your arms, sticking your tongue out and generally not behaving in the accepted way a 'maths teacher'—or a mum—is thought to behave. Before you play with number sequences, some pattern games without numbers give the idea.

First you create a pattern, then you repeat it to your child a couple of times before asking him to continue in the same way. These can be word patterns or physical ones: hop, skip, jump. Or clap, stamp, wave.

You could try pointing to your nose, your tongue and your ear, then beginning again with your nose, your tongue and...

You can use names of siblings, 'Susie, Edward, Joe.' Your child will love games using his own name.

The idea is to create an understanding of pattern, and have your child predict what comes next.

Other forms of pattern may use coloured beads or bricks. Try stacking the bricks in a particular order of colours, then offer your pattern-maker the rest of the bricks to continue.

Colouring in small shapes in a colour sequence will produce the same understanding: Draw a row of small circles with a stencil. Colour the first four red, green, blue, yellow. Ask your child what colour should come next.

## Sounds

Sounds can form patterns, too (but only for days when you feel invincible). Prepare a row of objects to hit. Good ones are cake tins, pots and pans, a plastic bottle or a baking sheet.

These are your drums, and you tap or strike them with wooden or metal spoons. The pattern is not only made of the different sounds produced when you strike the various 'drums', but rhythm, too plays a part. Rat-a-tat-tat...Can your little drummer reproduce the sound you made exactly?

You might have some fun making a loo roll snake or a cotton-reel caterpillar. Simply collect the cardboard cylinders of used loo rolls, paint them and thread them onto string. Empty cottonreels can be threaded onto a length of ribbon or wool. Make faces and antennae for these creatures from cardboard and two pipe cleaners. Necklaces made from macaroni are also popular.

All these suggestions allow you to compare lengths and patterns. Sequences will come to be very important, and these threaded-up toys can be done in a pattern of colours or shapes—two red ones, one blue; long cottonreels and short ones, etc.

We can introduce our children to the multitude of patterns in nature and in design. There are patterns, sequences and relationships to be examined all around us.

## Prints and Rubbings

Making prints of objects is to look more closely at their patterns. Rubbings in turn offer a very tactile experience. Put a coin beneath a sheet of paper and invite your child to rub over it with a wax crayon. An imprint of the coin will appear as if by magic. Dip a wooden brick into dissolved powder paint and water mix and press it onto paper. A perfect print of its shape appears. Do this with other bricks or your handprints and see if any are the same. Rubbings and prints of leaves are fascinating and offer many opportunities to study shape and form. (Rubbings also help your child understand pattern and symmetry—as we'll see later in this chapter.)

### Rubbings and Prints of Patterns

One way of emphasizing and enjoying a pattern is to make a rubbing. We have made rubbings of the designs carved into our diningroom chairs, the Spanish floor tiles, manhole and drain covers, flowers, leaves, and any textured surface with a pattern.

Rubbings can be made using wax crayons but it is worth investing in bars made for this purpose at art shops. They last for many rubbings longer than wax crayons and are very fat but less greasy and easier to use. They produce a better result. Simply lay your sheet of paper over the surface and rub evenly with the bar (or crayon).

We make prints of other types of patterns, dipping daisies into paint, experimenting with dried flowers, corks, potato shapes, organizing them on the page into another pattern at will.

Introduce young children to symmetry. Try to increase their awareness through observation and experience. The first form of symmetry to look at is simple. The shape is equally reflected on each of two sides of a line, called an axis. Look at a picture of a butterfly, for example. Then make some butterfly blob pictures using a fold as your axis.

### Butterfly Blobs

Have your child make a thick blob of paint in the middle of a piece of paper. Fold the paper accurately in half, pressing onto the paint. When you open the sheet you will have two symmetrical images. The blobs of colour will be reflections of one another. Try painting one half of a butterfly wing and printing the other half by pressing paper onto the first one.

Cut a fruit in half down the middle and examine the two halves. Search for things you can find that are symmetrical.

In your search you might come across some letters of the alphabet. The two halves must always fit each other exactly.

There are more advanced pattern and symmetry projects in Chapter 8.

## LANGUAGE FOR LEARNING AND THINKING

Chatting together at home or on outings you will find yourself endlessly trying to satisfy your child's curiosity as he seeks to understand the world around him. As you talk, make sure he understands useful concepts which the first-year reception teacher hopes he will know when he starts school. Is he clear on fewer than/more than, long(er)/short(er)/as long as, round/flat, curved/straight? Treasure hunts in which he must find things with curved surfaces, for example, are an intriguing and enjoyable way to help him learn.

## Word Collection

Enliven the odd boring minute making word collections. How many words can you think of that tell us something is small? Little, tiny, teeny-weeny, dwarf...
And big? Huge, gigantic, enormous, mountainous...
Hunt for opposites: wide/narrow, fat/thin, straight/curved.

Matching for size you can compare and use language to describe: 'Is this one longer or shorter? Is it as long as, or shorter than the other?' Drinking straws or old shoe laces can be cut up and used for this.

With your child, feel and talk about: shiny, smooth, slippery, slimy, rough, fluffy and sharp.

Describe the shape and texture of things—the postbox is a red, shiny, smooth cylinder. Teddy is a soft, fluffy, curved shape.

'The window is a rectangle.' Many children will say this at first. But look more carefully: it is really a thin cuboid. It is transparent (see-through) with a smooth frame. The pot lid is round, shiny and black, the edges are sharp.

## UFOs

These are only unidentified objects while the players are guessing what you have flying through your mind. Rather like 'I spy', the players have to guess what the invisible object is. You can help them by describing it and they may ask one question each about its shape or properties. You begin by giving a very brief description. 'It's a cylinder.' Then each player may ask a question. 'Is it large?' Answer only yes or no. Eventually they narrow it down through the questions and answers and guess your UFO.

Congruent means the same size and shape as another object. Children enjoy finding 'sames' or 'snaps'. But their skill with difficult words such as dinosaur names shows that they are more than capable of picking up words like congruent or cylinder. Check that your child knows words such as: alike, similar, different, another, corners, sides.

## Position Words

We cannot think without words, they are the tools of learning. Young children learn an extraordinary number of words in their first five years. Some estimates put this at a few thousand. Yet a surprising number of children in their first year of school do not fully understand these terms:

- next to
- above, below
- in front of, behind
- up and down
- beside
- under, on top of
- inside.

These are easily explained in an afternoon and would greatly help your child. Hiding inside, beneath or behind things is fun. Looking to see what is inside a parcel is tempting; arranging a table of objects and asking your child to pass you the one on top of the book, or in front of the jug, may be all that is needed.

## Arrows

We rely on arrows and the messages they give us almost without realizing it. Make it a game to hunt for arrows with your child, and check which way the arrows are pointing. It may be 'This Way Up' on a carton, it may be Emergency Exit at the theatre or something as basic as the toilet sign in a restaurant. Basic but vital. Handling these instructions and knowing what they say gives your child confidence.

If you draw a plan of your home, your child can draw arrows from the front door to his bed. On a map he can draw arrows to show where he lives, goes to school or visits a friend.

If Paul is new at school and nervous, and his teacher tells him to stand behind Tom, it helps if he can confidently do this. If she asks him to explore some mathematical concepts with apparatus he will need language to understand what she is asking.

'The bird is up in the tree and the cat is down on the ground,' offers adults a chance to say the words we use to label these positions. 'What is below the picture of the sea?' These terms can be woven seamlessly into your day, but make certain that your child is in no doubt about their meaning. Position is an interesting idea to think about. Is the object the same wherever it is? Is it still there even when you can't see it, behind something? Is it the same size when it is in a different position? Many children think it is not.

Help your child to think about your home. Whose room is next to his? Whose room is above the kitchen? Could he draw (with your help) a diagram and talk about this?

## Simon Says

A game of 'Simon Says' can be played in which you give instructions to players to stand next to someone or something, or put their hands above their heads, behind their backs, beneath their toes and all the variants you can think of and children can twist into.

# 3

• • • • •

# Making Friends with Numbers

## EXPERIENCE BEFORE SYMBOLS

There is an argument that teaching children true values before abstract numerals makes later maths learning much easier. Basically what this means is that you cannot understand a label or a symbol for a number, such as **2**, until you have met two objects forming a pair so often that you know what all pairs have in common. So what does this mean for a parent? In a nutshell, don't concentrate on teaching your child the names of the numbers in order (rote learning) until she has an understanding of their meaning. Wait until your child has plenty of experience of handling numbers of real things.

Symbols are essential in maths. We need them to manage and express concepts. The way we do this, our special system of representing quantities with symbols, has to be learned, but there is a path of progress. Understanding comes first in a series of steps. A baby experiences physical objects and then learns the language to describe the experience. Pictures are used to represent this experience: 'this is three' (see diagram on page 22). Finally, written symbols are an abstract representation of it: **3**.

So, first things first. The experience. Children do indeed start with experience and language on your living room floor or in the bath, kitchen or supermarket. You count aloud as you handle buttons, shoes, knives and forks and toys. Hopefully you'll

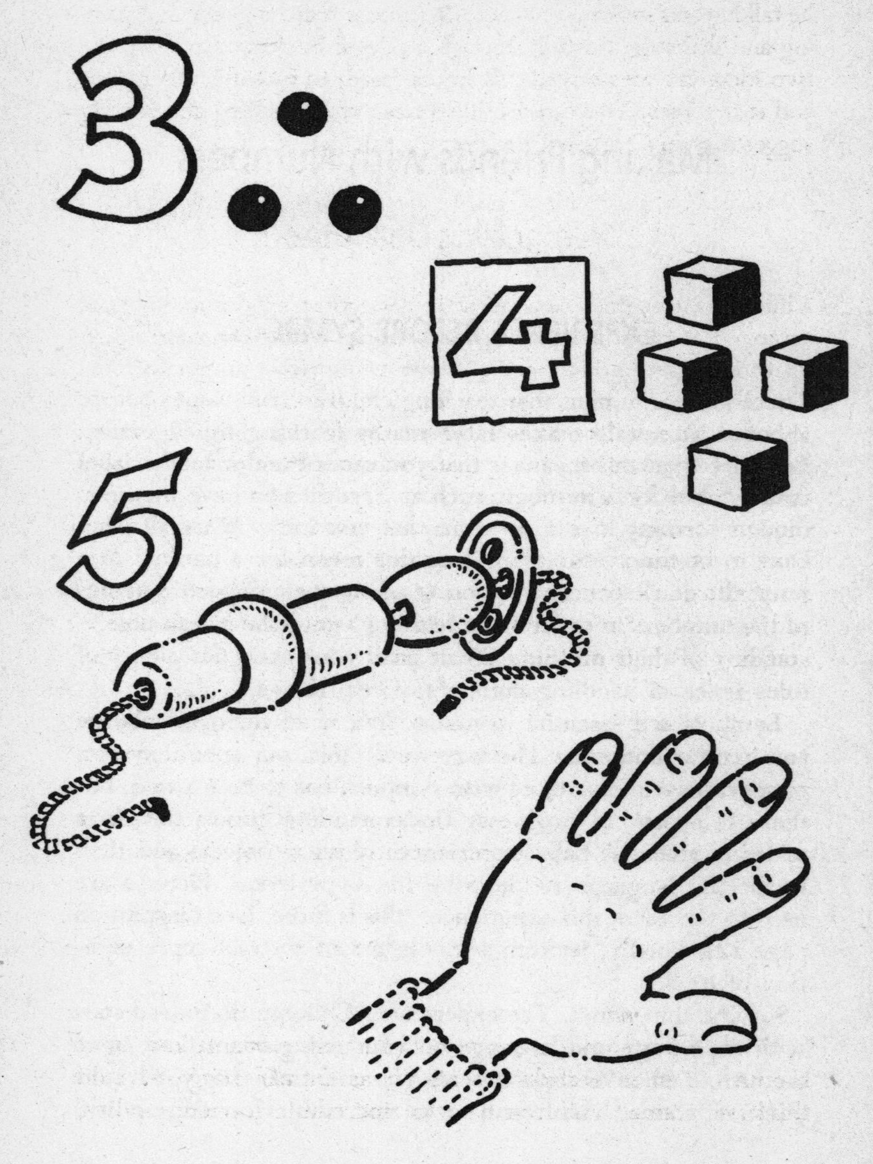
3
4
5

be talking and showing. Part of the game is matching, sorting, pairing and ordering. Sorting through a pile of bricks and putting the two identical ones together is to compare, to look for differences and similarities. You can use ideas from the matching and sorting suggestions in Chapter 2 for this.

## Recognizing Quantities

Children, it seems, can instantly recognize a number of dots arranged at random as a quantity, where adults can only do so up to about seven or eight without having to count them individually.

Some parents like to take advantage of this ability by using flashcards with dots on for a few minutes each day with very young babies. There are experts who recommend this method, though I must admit to never having used it personally. Glen Doman of the Institute of Human Potential argues that this gives 'small children a staggering advantage in learning to do and actually understand arithmetic'. He goes on to say that if you give a child the facts about maths, she will discover the rules.

Looking at a numeral, which is simply a symbol, a label if you like for a number of things, is not as clear as looking at a group of dots, or handling some objects yourself. Between the ages of one and two, Doman believes that babies can look at an arrangement like these and instantly recognize the quantity.

••• + ••• = ••••••

While I cannot imagine myself flashing cards at my eight month old in several daily sessions, as Doman suggests, I do think very small children have an incredible learning ability, and it seems obvious that to understand numbers they need concrete experience rather than being taught symbols first.

If this makes sense to you, you can make cards or a frieze that includes dots as well as numerals. I would be happy for a child to play with the cards from the age of one onwards and at a later age to try to match the dots with other cards with numerals. I think it is a great help to the child if you tell her 'this is one (•), and this is how we write 1', thus distinguishing between true value and the numeral. But I'm not really happy to tell you to get going at eight months with a daily flashcard lesson.

This is not so much because I don't think the baby could take it, but because not all parents can remain relaxed when involved in an intense programme of this kind. True, their child might become brilliant at it, but is their relationship on the line? A tired working parent might not have the inner resources to stay unfussed if the child doesn't want to co-operate. After work may be the worst time for your child. A few cuddles and a storybook might be the best thing at this time. I prefer to avoid a rigid timetable. The child's early years have been described as a timeless dreamtime and I like to think of them like this; there's plenty of time later to be corralled into schedules and timetables.

I prefer to put energy into talking about and playing with concrete items such as big beads or wooden bricks. You can build towers and count how many bricks you can pile up before they fall down. You can thread beads onto a string and talk about how many there are. You can count the buttons on her shirt or jacket as you do them up. You can simply bring numbers into your child's life at a young age because they are everywhere around you. I have never set up sessions, fixed pre-arranged learning times with small babies and infants. I am happiest going along with their mood and the exciting opportunities for learning that the average day throws up. Letting yourself be led by the baby reveals the remarkable way in which she is learning. This way you can show her relevant new ideas that fit in with the direction she chooses. It seems less sterile and cold than cards.

In fact, I believe that one of our problems is that we don't let our children lead and discover for themselves as often as we could. This is probably because I am a parent and not a professional educator. Of course this means you have to devote a great deal of time to just being together, and observing your child, companionably being alongside and always watching for safety. This is one of the great pleasures of parenting, and should be free of guilt about what you should be doing instead, be it housework or using flashcards.

Then there is answering your child's questions and encouraging her to think and ask even more. A child may repeatedly practise a new skill until she has mastered it. Nothing else interests her while she is doing this. She keeps returning to it. She may also keep asking you about something she is thinking about. Now is the time to explore that idea more fully, because she has shown that *she wants to*.

Curiosity may be mistaken for lack of concentration. If your child will not concentrate on what you are trying to have her see, learn or do, does she have some other absorbing idea taking her whole concentration and drawing her ever back to it? Life may become easier for both of you if you go with the flow and follow her enthusiasms. There are always opportunities to introduce numbers in any form.

Discovery can be helped a little by making it more likely. You can certainly help your child meet the world of numbers early on and encourage a fascination. You can have up your sleeve a mental 'menu' of things you have read about in a book like this, or remember enjoying as a child yourself. You can read up on a stage of development and be ready to lead your child through the next doorway on this journey. I am not convinced, though, that parents should feel guilty if they are not flashing cards at their infants in a set programme.

## COUNTING

Many young children will happily recite 'One, two, three...' if you ask them to count. But saying the names of the numbers and knowing what they represent in concrete terms are not the same thing. Reciting the numbers can be a party trick and is usually met with our admiration. But asking this adorable reciting little person to give you a number of things may reveal that she has no idea yet of what counting is for. She may also have no knowledge of what the number symbols look like.

Researchers have identified three stages that children go through when learning to count. The first is the rote counting I have described above. A child can recite the number words in the order you have taught her, rather like a nursery rhyme. How does this child imagine the numbers? As a procession of imaginary creatures like ET or The Smurfs, jostling for their position in the line?

In the second stage a child understands the necessity of naming a number for every object counted. She understands that the order in which objects are counted will not affect the total. She knows too that the last number counted labels the group. This is a big development from the child who believes that when we count three cups, we are giving a name label to each cup.

This child looks on as you count, pointing to the cups: 'One, two, three.'

She will agree that you have three. But when you take a cup away and ask 'How many cups are there?' she will say three. When you remind her that you have taken one away, she will agree, but argue that you still have two and three. She has not realized that the last number describes a property of the set as a whole, rather than the last object. She has not met numbers as adjectives first rather than nouns.

Understanding is easier if a child first meets numbers as

'two eyes in your face, three coins, four legs on a dog'. Handling real objects makes it clear. Touching and counting each object makes the one-to-one correspondence clear. If you use pebbles or bricks, touch and move each one to the side as you count them.

Play with dominoes or matching cards with dots helps practise recognizing a number of dots. You can make your own set of Domino cards according to the instructions on page **28**.

An alternative to this is to use dice. Throw a die and ask your child to give you the exact number of pebbles or paper clips shown on the die.

In the third step, the child learns to conserve number and understands the permanent value of what she counts. Adults know that if we have a packet of 10 sweets, they remain 10 sweets whether they are in the packet or spilled onto the table. Children do not automatically know this. We know that the quantity of something is conserved even if it changes shape. Some children perceive that when an object is moved it becomes a different object. This understanding about the 'threeness' of three may come sometime during your child's first year at school.

So, to remind ourselves about the stages of counting:

- first there is simply saying the numbers in order
- then there is making a relationship between the counting numbers and objects on a one-to-one basis
- finally there is the ability to conserve.

Some educators would prefer to leave out the first rote learning stage and have children learn by discovery about numbers as describing words. A child is shown one crayon, two bottle tops and three coins. Gradually through play and talking she will come to understand that two is one more than one, three is one more than two. She can then arrange them in order.

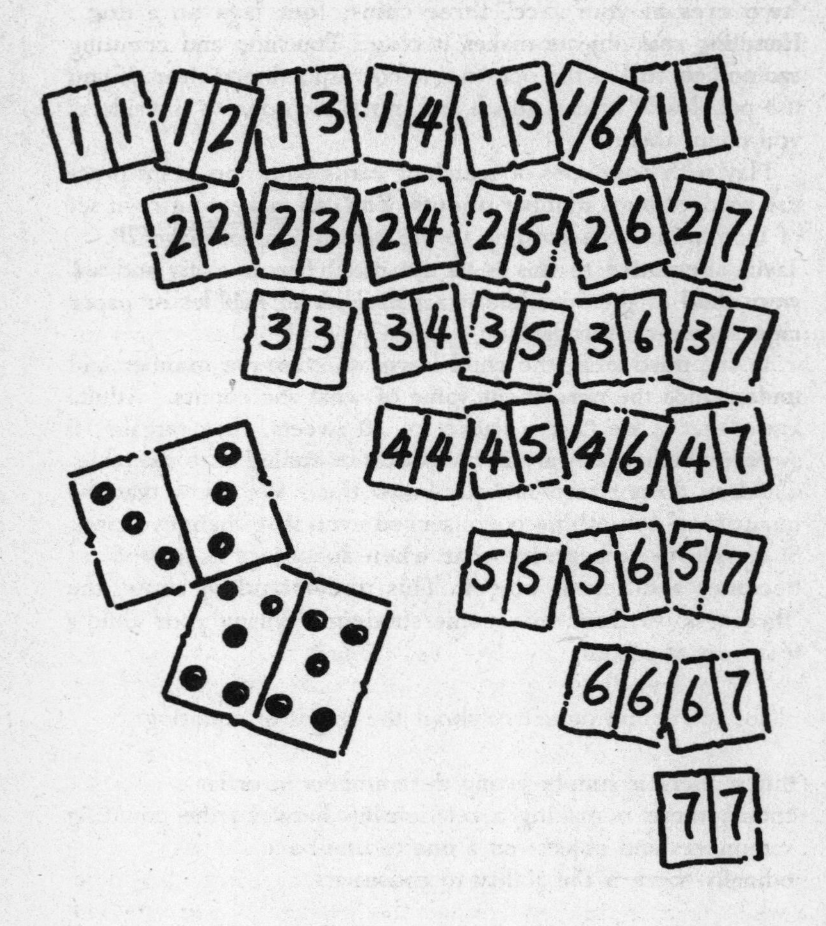


***Make your own set of domino cards. You will need a set of 28 stiff white cards—if coated in plastic they'll withstand handling better. Numbers can be drawn on, or you can use dots set out in any arrangement you wish.***

She then learns the adjectives we use to describe these quantities, and should grasp the reasoning behind the order of one two and three.

## Personalized Rhymes

Having your own name in the story, rhyme or song pleases every child. Make up your own special versions of well-known ones or invent another. You can easily develop these verses by bringing counting into them. They can be funny or rude; small poets love to include family jokes:

> There are five cookies on the dish,
> Soon Simon'll get his wish,
> He'll take a cookie and run for the door
> And on the plate there'll be only four.
> There are four cookies on the dish
> Soon Sam will get her wish,
> She'll take a cookie and climb a tree,
> And on the plate there'll be only three.

Children should also meet numbers in an arbitrary order sometimes, so you might introduce two eyes and five fingers, for example. Numbers are not always met in a fixed order. Our bodies are evidence of Nature's random arrangements. It will be a while before a child understands the 'fiveness' of five. She will then grasp the use of the word five as a noun, indicating a quantity of five. Games such as 'Can you find me a five?' (Answer: my toes) will develop.

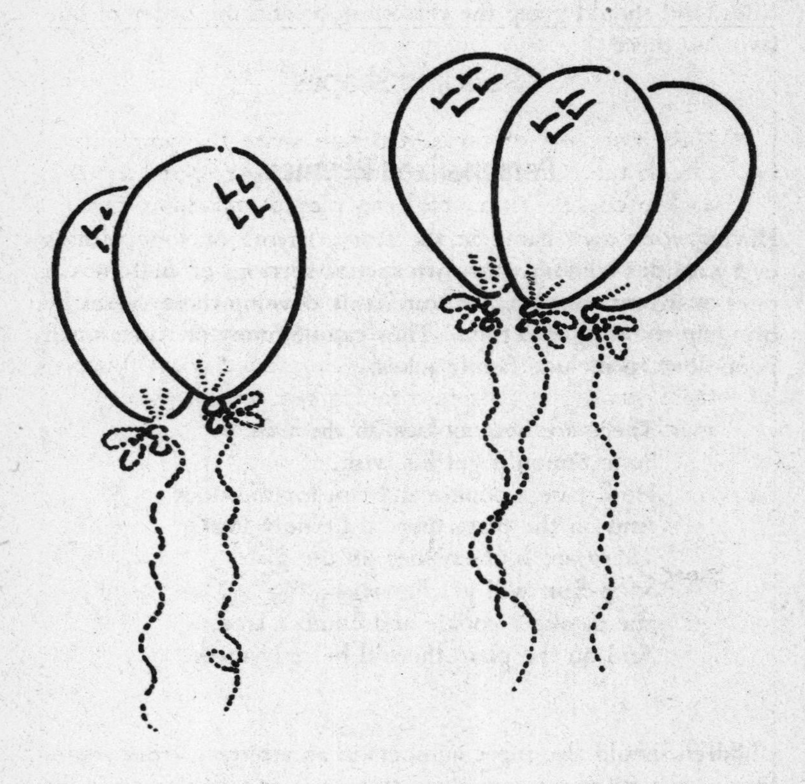

*Five balloons*

As well as learning to say the names of the numbers, a child will learn to recognize the symbols we use for each number, and to draw them herself. There should be no rush about this. It is far more vital to have a good understanding of handling numbers of concrete things first.

## GETTING TO KNOW THE SHAPES OF NUMBERS

### Dots and Shapes

- Make two sets of cards. On one write the numerals one to ten. On the other mark dots corresponding to each numeral. Your child can play at matching these up.
- Cut out number shapes from a wide range of materials with interesting textures. Try a choice of silky satin, rough sandpaper, tin foil, hessian, smooth card or silver/gold paper saved from chocolates. Have your child stroke the shapes, running her fingers over the numbers and becoming familiar with them.

*Textured shapes are fun to work with*

- Make a large poster. Draw a generous fat solid shape, i.e. **6**. Invite your child to colour in patterns or stick old scraps of wallpaper over it. Some children enjoy drawing groups of six objects all over it. Six hens up here, six fish down there. Six dots, six stripes...
- Put up a number frieze. You may buy these easily but they are also fun to make. Simply make/paint or draw clear number shapes, each in its own square. Within the square glue or draw the number of objects your number indicates. You may use old photographs, pictures torn from magazines, handprints, buttons sewn on...Put this where your child can see it easily.
- Make some cards with numbers from one to 10. Together (at first), show your child how to find the right number of objects–buttons, coins, tiddlywinks, bricks, wooden beads, etc. and match them to the right number card.
  Now invite your child to do this alone.
- Try numbered jigsaw cards. For these you will need to use a fairly large card. Draw a number six on a card. Now draw six hens, cakes, buttons, etc. Cut the number away from its six objects by a zigzag cut. Repeat for all the numbers. Mix up the cards and invite your child to sort them and match them.

***Numbered jigsaw card***

- Use plasticene or playdough to make the shapes of the numbers. With these marvellous 3-D numbers you can play blindfold games to guess by touch what number your hand is feeling. Homemade dough recipes can be baked and glazed for long-lasting play (see Appendix 2). When she can recognize all the numbers blindfold, step things up a little. Turn the numbers over or put them the wrong way up. Can she still tell without looking? Now have her check her guess.
- Give your child a pile of bricks or blocks. Show her a number you have drawn or made. Ask her to build a tower using the same number of bricks.
- Draw numbers with a wet finger on a pathway, on a steamy window or a dusty ledge. Make the right number of finger prints next to the number.

## INTRODUCING COUNTING GAMES

- Invite your child to thread buttons or beads onto a shoelace. Ask, 'How many buttons have you threaded? Can you point to a number which says this?'
- Ask your child to consider how many legs people have. How many do animals have? How many legs do spiders have? With plasticene or playdough model some creatures with emphasis on their legs. Invite your child to make you a series of animals with four legs, or eight legs. Perhaps you might link this to a story about a dog or a spider.
- Hand and footprints. On a hot day this can be an outdoor activity. If indoors lay a plastic sheet or old shower curtain/plastic tablecloth on the floor. Prepare some powder paint and water, making it up fairly dry. Add a squirt of washing-up liquid. Put it into a flat swissroll tin (these are so valuable for artwork of many kinds, hunt down an old battered one and keep it handy).

Lay large sheets of paper or a roll of wallpaper lining paper on the floor. Ask your child to remove her shoes and socks, and wear as little as possible. She may dip a hand or a foot into the paint and then step or press it onto the waiting paper. *But* there are only to be a certain number of prints. Choose a number to match your child's counting skill and see if she can make the correct number of prints.

Hose off surplus paint. Leave prints to dry. Hang up the work where you can see it and chat about it. Look at toes or fingers. Are there other things in the picture you could count together?

- Use counting games and rhymes with fingers and toes while in the bath or dressing. 'Hide' a couple away and ask your child to guess how many are left.

> This little piggy went to market,
> This little piggy stayed home.
> This little piggy had roast beef
> This little piggy had none.
> And this little piggy cried
> (squeal, squeal, squeal) all the way home.

How many little piggies are there?

- When waiting around somewhere with nothing to do, see how many people are wearing jeans. How many people are there on the bus? How many just got off? How many got on?
- Draw purses on paper. Put a few pennies into each one. Name the owner of each purse (use imaginary characters, teddies' names, friends, or family). Give your child another card on which you have written what each purse owner spends. To begin with, use very low numbers. Mum spends 3p.

Suggest that your child takes the money out of Mum's purse and count how much is left. Now do the same for all the others. Do you know how much was spent altogether? How much was left altogether?

## Dice Games (ages 4 and up)

Any dice games which have players moving along a track with each roll of the dice will give practice in seeing a number of dots and then moving a counter the same number of places. (Snakes and Ladders is usually recommended for this, but I find that some very young children get too distressed at losing when landing on a snake. Losing and staying in there has yet to be learned. Too much distress may put a child off numbers. Keep this game for seven year olds.) Draw a simple track on which you can only go forwards. The track itself may be wiggly or spiral or straight.

## Racetrack 1

Draw a few straight racetracks alongside each other with a ruler, ending in a finishing line.

Mark off squares of equal size all along the tracks and draw a starting line. Invite children to choose a little counter or toy to race. Place starters ready. Each player throws the dice in turn and moves the number of squares shown on the upward face of the dice. Who will cross the finishing line first?

As children grow older you can upgrade this game with extra instructions on the squares which read, 'Go back to the start/go back two squares/go forward two squares' etc. Racetrack 2 (see Chapter 5) is useful when they can add, subtract and multiply.

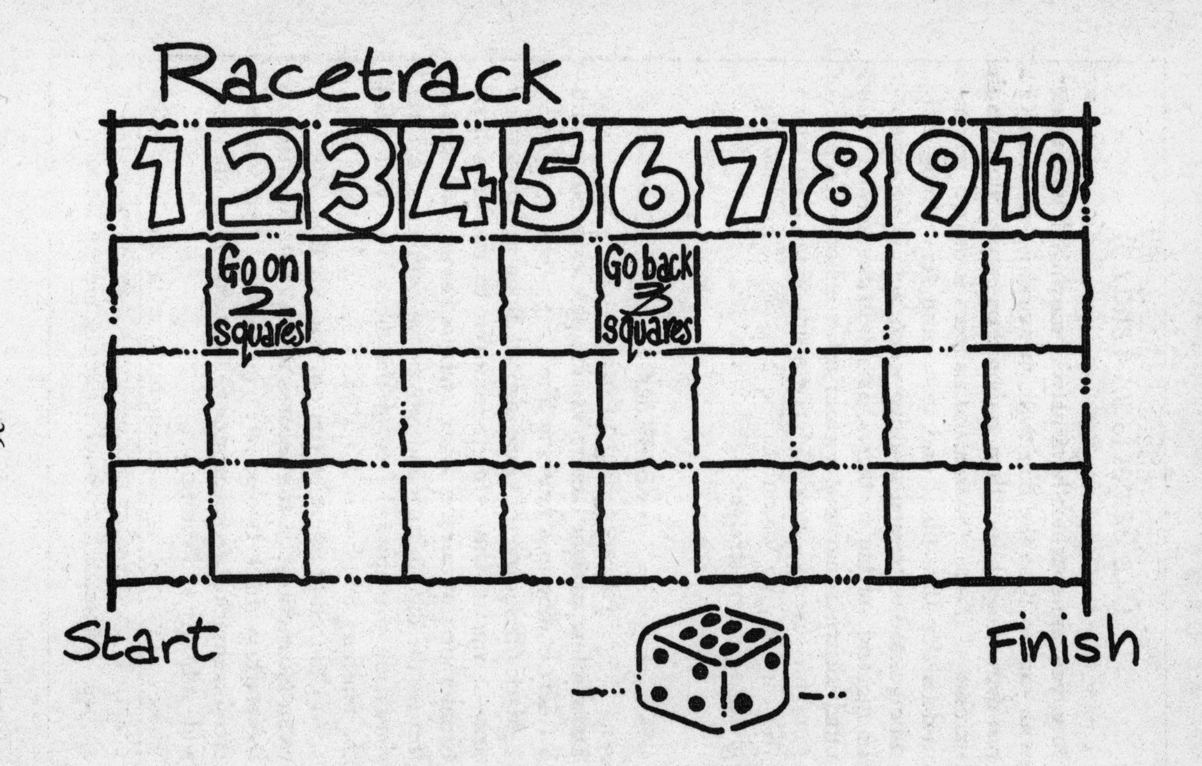


Racetrack 1. For very young players make a track on which you can only go forwards.

## USING NUMBERS IN REAL LIFE

Of course you will be using numbers in real life at the same time as you use specifically targeted games. Talk about numbers as you do the shopping together or sort the socks. Chat as you dress to go out...'I can only see one shoe, where is the other one?—There it is, under the bed, now you have two. Let's do up those buttons, one, two, three...they're all done.' Numbers make sense of these daily activities. Children understand their importance when you show them how much you need to use them, making numbers part of living.

### Teatime by Numbers

When preparing a tea party, talk about the number of sandwiches you are going to make. Give your child a packet of sliced bread and ask her to work out how many slices you are going to need. Have her count out cheese slices for this feast and possibly tomato slices or lettuce leaves which you have prepared. Why is the number of bread slices different from the number of cheese slices?

### Do a Survey

Who likes cheese? Let's count up these people...
Who likes orange juice?...
How many orange juices should I get then?

These useful findings can be written in the form of a chart. Dots or stars can be used instead of number shapes. This makes it easier to add someone later.

## Place Settings

Suggest your child paints a few dots on paper napkins, and place them around the table, one in each place. When the children come to the table you tell each one that his or her place is to be found in a secret message. Give each child a folded piece of paper on which you have written a number to correspond with a place at the table. They open their secret messages and work out where to sit.

You can carry this familiarity with numbers over into your child's fantasy life, which is no less real to her. Invite her to set out a picnic for toys or imaginary friends from stories. Ask your child who is coming to this 'play' tea party. How many plates, cups and paper napkins will be needed?

Liane and Louise planned a wedding for their Cabbage Patch dolls. Elaborate arrangements were made. Vital journeys were made between our house and Louise's. Were the teddies invited? Who would be the vicar? How many were coming to the wedding tea? How many tiny biscuits should we make? How many sandwiches would we need? Chairs, cups and saucers from a miniature set, music and confetti completed the day. A wedding dress was made (how much fabric should we use? Is this remnant big enough?) The groom got a new suit made on the dining room table. An old shirt was cut up to provide his impeccable striped shirt. Black oilcloth made shiny smart shoes.

Neighbour Toby loaded up his train set with the toys who were to attend the wedding. How many toys fitted into each carriage? How many carriages did he need to use? These were real problems to be solved. Would he have to go back and make another trip? The discussion was valid in his eyes because he was considering the problem himself.

Louise's mum and I did not set out to 'use numbers', only to have some fun and bring the girls' dream to life. But we needed numbers. This is the most useful lesson of all. We had problems to solve and answers to find, and the numbers helped us do this.

## Numbers Treasure Hunt

For two or more players. Give each child a treasure hunt list. They then go off and search for the items you've listed. If there are very young children playing, team them with older partners. You will have hidden the 'treasure' beforehand and decided which rooms of your home are out of bounds.

Your list may look something like this (if some cannot read in the group, use drawings):

FIND and put into this paper bag:

- 2 pennies
- 1 shoelace
- 3 hairclips
- 5 red sweets.

When they have found everything on the list, they rush to you to have their bags checked. You may need a helper at this point to check them all.

If the question 'Why should I learn about numbers?' is lurking about in your child's mind, the answer should begin to emerge from the way numbers seem to play such an important part in everyday life. It becomes clear to the child that knowing about numbers is not only intensely interesting but also useful. It is clearly a grown-up thing to be able to do. It makes sense of things. Working things out and solving problems is exciting and satisfying.

# 4

• • • • •

# Adding and Subtracting

No, it's not a major hurdle, waiting to trip your child up. It should be making sense already, and be a natural step after all those counting activities.

Adding will have developed naturally in your shopping, cooking and playing talk. There it is in everyday life, not something set apart and shrouded in mystery.

'We've got one of your shoes, where's the other one?' 'Here it is, now we have two shoes.'

'I've only got three cups, now that Susie has come we're going to need four, I'll go in and get another one.'

The easiest way to understand addition is to use concrete objects and your own two hands. Try to give your child as much play with several similar objects as you can. Talk as you pass him a few more, and ask him to tell you how many he has now.

## How Many Ways Can We Arrange These?

Give your child a number of playthings. These might be pebbles, bottle tops, counters or wooden beads. Hand him two cups. Ask him to see how many different ways he can arrange the pebbles in the two cups.

If the number of items is eight, for example, he should come up with six in one cup and two in the other, or seven in one and one in the other. Help to write out his solutions if needed. Each of these solutions shows ways in which numbers add up to eight. We can explore physically all the simple adding facts this way. Begin with numbers 10 and under.

## Magic Beans

Hold a few kidney beans in your hand. Have your child look at them and count them. Now he closes his eyes and you quickly and silently add a few more. He opens his eyes and looks at the number of beans in your hand. 'How did I do this?' you ask. 'I had two, now there are six!' Counting carefully he will work out that you added four.

Later, of course, you try this out with taking some away—Vanishing Beans. Children need to repeat and consolidate, often practising over and over again something you were sure they knew weeks ago. Reverse roles. Now he holds the beans and you close your eyes and he adds or subtracts from the total. You will now 'guess' how many he has added or taken away to get the new total.

## NUMBER LINES

With a number line, like the one below, your child can add and subtract by counting along the line.

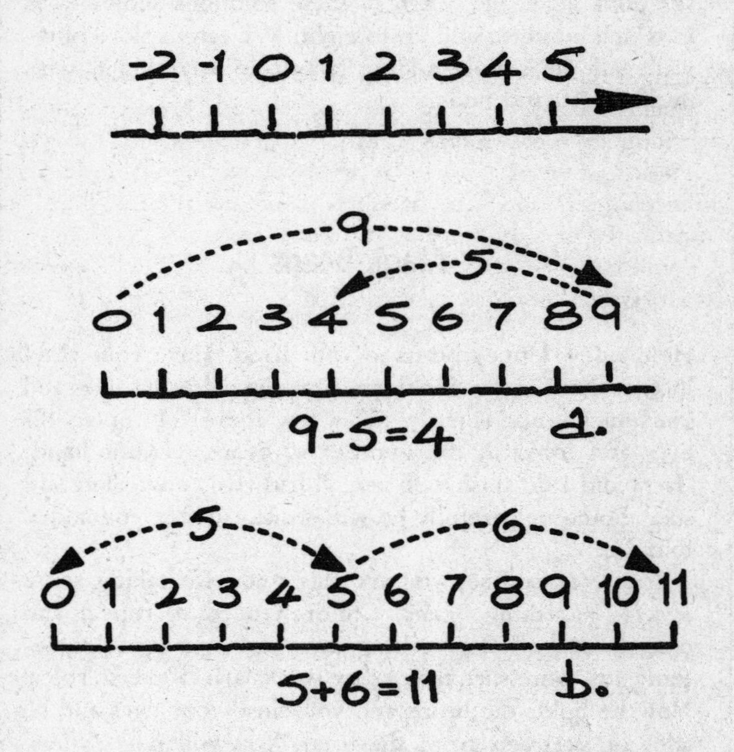


*Number lines: a. subtraction on a number line: taking 5 from 9; b. adding 5 and 6.*

Use your number line to show that subtracting undoes what addition has done:

**6 + 3 = 9, but 9 -3 = 6.**

Children can explore different ways of expressing the same number fact:

$$6 + 3 = 9, 3 + 6 = 9, 9 - 3 = 6, 9 - 6 = 3.$$

These are known as a number bond.

Making a number line on a piece of card can be a great help to your child. For a child who seems ready, you can include the numbers 0 and minus 1, etc.

These number lines can be used for adding—simply place three counters and four counters along the number line and you can see at a glance that you have seven.

Some children will simply count along the line backwards and forwards to add or subtract.

## Who Needs a Calculator?

For some speedy calculations try adding with two number lines. In the diagram below, the first three numbers of line a) are on the left, and the six numbers we are going to add to these three are on line b). How much is three plus four? See how the numbers have been arranged in the diagram: line b) is placed after the first three numbers on line a), so simply by reading along line b) for four numbers we can see (looking up to line a) that the answer is seven.

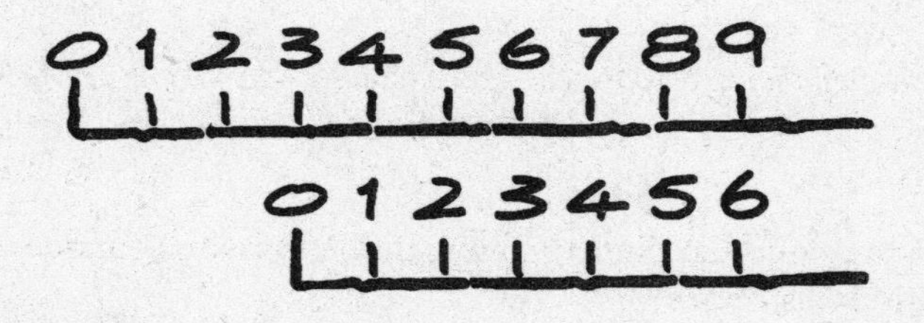


*Adding with number lines*

## Dice Practice

Deciding who shall go first in any game can be decided by a throw of the dice. He who has the highest score goes first. To find this out your child will have to add together your score (the result of throwing a pair of dice) and his score and see which is the highest.

## Who Goes First?

In all games played with dice the importance of throwing a six needs to be understood. If you throw a six you get a second turn. You will need to add together both the six and the score from the second throw.

Although many children will always do their adding and subtracting by counting, being familiar with these number facts through play can lead to them simply knowing that six take away four is two because they have done it so often with pebbles. This speeds things up for them and gives them confidence.

For further games using a calculator, see p.91, plus crossnumber, sell-by date and equals, pp.92-93.

## Build 10

Playing this game practises making 10 by addition. Draw a hexagon like this:

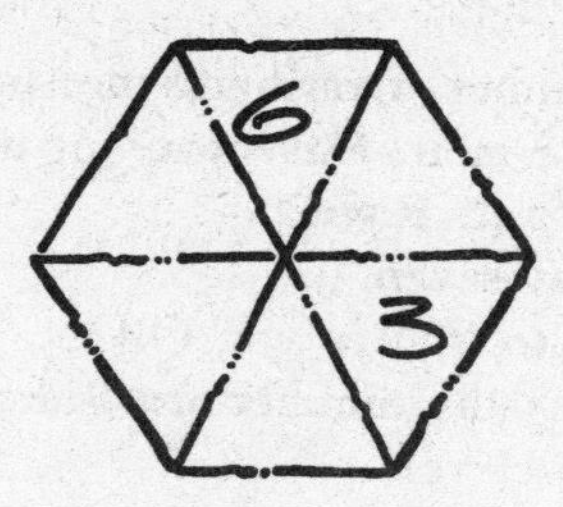


Numbers in opposite segments equal 10. Invite your child to complete them. You could add the numbers 4 (6 + 4 = 10), 7 and 3, and 8 and 2 to this one, for example. Change the total to try it with 12, or eight.

Gather some counters, tiddlywinks, pebbles, etc.

Ask the children to find ways of placing two, three or four counters on the numbered hexagon so that the numbers covered add up to 10/12/20.

When you are stuck in a traffic jam or a queue, you can play games with your fingers. Hold up all five and check that your child agrees on how many there are. Then you hide two: 'I've taken two away, how many do I have left?' Gradually build to using two hands. As his skill grows, he will soon be able to do this game in his head without seeing your fingers at all. You'll be able to say, 'I've taken three away, how many do I have left?' He will be familiar with adding and taking away up to 10.

## Think of a Number

Children love to try this puzzle on their family, friends—anyone who'll join in.

- Think of a number. Don't tell me, but remember it.
- Now add three to it. Have you got it?
- Now add 10 to it. Right?
- Now take away seven.
- Now add five to it. Are you OK?
- Now take away the number you started with—
- The answer is eleven.

# HUNDREDS, TENS AND UNITS

No child is content for long to add or subtract numbers that add up to less than 10. When you were playing the counting games mentioned in the last chapter you will have occasionally counted in tens. You need to remind your child of this and show him that in our number system we count in tens. We choose a position for a number which tells us what its value is. For example, if we were to write 22, the place in which we have put the first 2 tells us that it is the symbol for 2 tens. The second 2 is the symbol in this position for 2 units. If we write 222, we have 2 hundreds, 2 tens and 2 units. So the place where the number is arranged tells us about its value. This is known in maths as place value.

## Base Ten and Place Value Games

You'll need three containers—cottage cheese pots will be fine. Collect all or some of these: a couple of hundred matches, a pile of kidney beans, tiddlywinks.

- Label one container 'Hundreds', the next 'tens' and the third 'ones'. Arrange them on the floor with the hundreds on the left, the tens in centre and the ones on the right.

  Make a starting line within a reasonable distance from the three pots arranged in line on the floor. Give each player 10 beans. Take it in turns to throw your beans, one at a time, towards the pots. Any that don't make it are worth zero, unless the children playing are very young; to keep them happy simply gather up their missed throws and give them another go. The object is to score as high as you can. At the end of each player's turn, count up his scores for each container (that is, three beans in the hundreds pot, five in the tens box, two in the units box) and write it down on a sheet of paper. Prepare this sheet with a triple box for the total: a box for the hundreds, a box for the tens and a box for the ones. Chat about the position or place of each digit and the value that place gives to a number.

- Start another game by arranging your three pots in the same way as above. Gather a large number of matches. Select ten single ones ones and show your child how to replace them with one match in the tens pot. Put the bundle of ten aside. Now carry on counting matches and sorting bundles of ten. When you have ten matches in the middle pot it is time to bundle them up and replace them with one match in the hundreds pot. Put the bundle aside. Carry on until the matches are all counted. If you wish to check your count ing, the neat bundles of ten you set aside make this easy—provided you have placed them in the same positions: hundreds on the left, tens in the middle, ones on the right. This game shows physically how our number system is based on tens, and also demonstrates place value, the idea that the value of a number is shown by its place.

- Another game to show place value is played with a pair of dice. Each player rolls the dice and reads his numbers. Then the player must decide how to write them down to produce the highest score. If he gets 5 and 3, should he write down 35 or 53? After several rounds of this, try it with three dice. The winner is the player with the highest score.

Children often appear to understand this perfectly for months on end and then suddenly need a new explanation all over again when they meet some new aspect of maths. Playing with bundles of matches and pots can make this clear. Take 15 matches. Count out ten, and put an elastic band round the bundle. Place this bundle one place to the left in a pot. This represents one ten. Five more matches remain. These are units. You have one ten and five units, written 15.

Now, using 115 matches, have your child count out bundles of ten and use an elastic band for each one. When 10 bundles are reached they are tied together with a big elastic band and moved one place to the left, representing 10 bundles of 10 = 100. One bundle remains in the position of the tens, and five matches remain in the units section. This tells us that we've got one hundred, one ten and five units.

We write addition down this way:

$$\begin{array}{r} 8 \\ +4 \\ =12 \end{array}$$

We added the 8 to the 4 and we got 12. 12 is one ten and two units, so we carry the one ten over to the left (to the tens column) and write the 2 in the units column. To begin with it helps to head columns with a word or number that reminds your child which is which.

There are many types of mathematical apparatus available, such as an abacus or rods. Schools will be likely to have these, but you can improvise at home. Many children have an abacus, but you can use any vertical stand and drop rings onto the uprights. Each time you get 10, you replace them with one on the next upright to the left. These may be rods from a construction set, or nails in a piece of wood with old-fashioned brass curtain rings dropped onto them.

## Patterns in Addition

Draw a grid like the one shown below. Help your child to add the vertical and horizontal numbers together—but with a twist: instead of writing the answer in the box, draw a star there if it is, for example, 11, a face if the answer is 13, a flower if the answer is 9, or perhaps colour the box in red if the answer is 7. What kind of patterns are you left with at the end of the game?

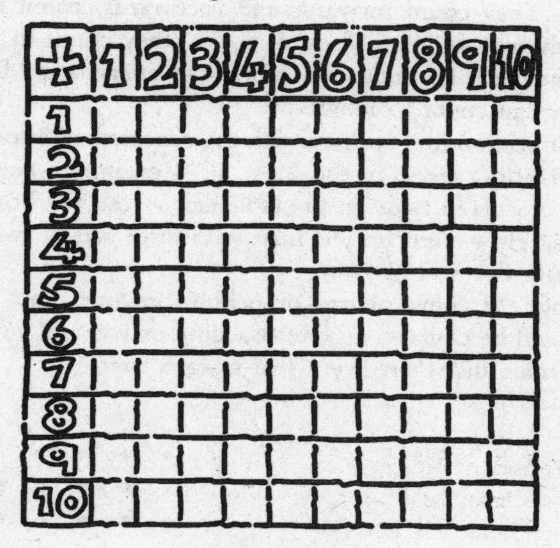


*Patterns in addition*

5

. . . . .

# Multiplying and Dividing

Children first meet multiplying without knowing it when they are in the first excitement of counting mania. At this point they are in love with counting. They finely hone their skill. They count for fun to see how far they can get. They count to show off to you and themselves how good they have become. They count forwards and backwards, count in twos and in threes—2, 4, 6, 8 or 3, 6, 9—they count in tens or fives. The game becomes ever more sophisticated and they tell you they can count to infinity!

Multiplication first appears with the repeated addition of the same quantity, the 2 in the 2, 4, 6, 8 example. You might arrange objects in twos on the table and investigate how many you have. How many do you have with three sets of two? How many with five sets of two?

Arrange 20 coins, acorns or pebbles in four rows of five. They could be counted up one at a time or you and your child could notice that there were five in each row,

5 + 5 + 5 + 5,

● ● ● ● ●
● ● ● ● ●
● ● ● ● ●
● ● ● ● ●

four times. Another way of saying this is 4 X 5.

## Number Circles

For this you need a die, large sheets of paper, gummed sticky shapes (all the same shape) and two felt-tip pens. Have your child roll the die. Whatever number comes up tells her how many circles to draw on her sheet of paper. Your turn now. You throw the die and draw your circles.

The second round determines how many sticky gummed shapes you may put into each circle. If your first number was three, you will have drawn three circles. If the second number you come up with on a throw of the die is six, you stick six gummed shapes into each of your circles. You now have three groups of six.

Add up your totals. The winner has the highest score. If gummed shapes are not available, draw in dots, stars or flowers, etc.

Many windows have small rectangular panes. We have fun looking at them. There are three across and four down. We count the panes. We talk about how three rows of four are twelve. Four rows of three are also twelve. With real objects, understanding is built up and your child will gradually be able to think about concepts in more abstract terms.

Playing Number Circles prepares your child for multiplying. Although you are not yet using the X sign, your child is becoming familiar with situations where she needs to find out the sum of three lots of five for example and she sees it visually. Multiplication groups numbers; if they are physically grouped into circles it is that much easier for your child to see this.

## The Three Bears Revisited

Going back to the toys' tea party idea in counting (Chapter 3), develop this story or build on another favourite. Talk about these bears and their tea. Draw a plate for each bear. How many biscuits do you think each bear might have? Bears love biscuits and after some discussion you and your child might decide that they'll gobble up at least five each. On each plate have your child draw five biscuits. Now you have three lots of five. How many biscuits will the bears eat all together?

## The Giant's Castle

The giant is having his friends over to his castle. Each giant is going to eat one melon and three enormous pies. Each giant takes his shoes off and leaves them by the door. There are four giants. How many melons will they eat? How many pies? How many shoes are left by the door?

## Nuts at the Zoo

Four children are going to the zoo. They each want to take a packet of peanuts to feed to the monkeys. Mum buys the packets at the gate. In each packet there are 10 peanuts. At the first cage the monkeys are so funny, they swing by their tails and beg for nuts. The children give them all the nuts. How many nuts did the monkeys eat?

## Automatic Counting

Children enjoy the confidence of knowing the answer before you demonstrate it. They enjoy even more knowing the answer before the calculator can prove it to them. When your child can add with confidence, you can show her that you are going to ask the machine to add in threes. Put 3 up on the display and ask her if she knows what the next number will be. She will say six and you then prove it on the calculator by pressing + 3 =. In this way add together or count in threes. Your child may use her fingers to add another three to the number on the display. Each time you add another three, draw a line on a sheet of paper to keep track of how many you have added including the first one. Only show her the answer on the calculator after she has predicted the answer. When you reach a number such as 27, stop and count up the number of strokes you have drawn. This will show that you have added three nine times. Explain to your child that now you are going to show her a quick trick to get this answer another way. You have added 3 nine times. So if you tell the machine to do this for you, you press three and then the sign for times (x), then 9 and =: up should come the answer 27. You can introduce the idea of times (x) this way.

## MULTIPLICATION TABLES

The easiest order to tackle them in seems to be 2, 5, 10, 4, 8, 3, 6, 9 and lastly 7. Of course 11 is so easy it needs only be shown once or twice. 12 can be learned later.

Charts of tables set them all out in an orderly fashion, and children seem to learn them less often by chanting these days, though old records such as Danny Kaye singing Inchworm may lurk in your attic.

## Make a Table of Tables

- Along the top line set out the numbers 1 through 10.
- Down the left-hand side, to the left of the 1 in the top line, set out another 1 through 10. (See diagram below.)

| | 1 | 2 | 3 | 4 | 5 | 6 | 7 | 8 | 9 | 10 |
|---|---|---|---|---|---|---|---|---|---|---|
| 1 | 1 | 2 | 3 | 4 | 5 | 6 | 7 | 8 | 9 | 10 |
| 2 | 2 | 4 | 6 | 8 | 10 | 12 | 14 | 16 | 18 | 20 |
| 3 | 3 | 6 | 9 | 12 | 15 | 18 | 21 | 24 | 27 | 30 |
| 4 | 4 | 8 | 12 | 16 | 20 | 24 | 28 | 32 | 36 | 40 |
| 5 | 5 | 10 | 15 | 20 | 25 | 30 | 35 | 40 | 45 | 50 |
| 6 | 6 | 12 | 18 | 24 | 30 | 36 | 42 | 48 | 54 | 60 |
| 7 | 7 | 14 | 21 | 28 | 35 | 40 | 49 | 56 | 63 | 70 |
| 8 | 8 | 16 | 24 | 32 | 40 | 48 | 56 | 64 | 72 | 80 |
| 9 | 9 | 18 | 27 | 36 | 45 | 54 | 63 | 72 | 81 | 90 |
| 10 | 10 | 20 | 30 | 40 | 50 | 60 | 70 | 80 | 90 | 100 |

*Table of tables*

There are some useful things to look at when the chart is filled out. The product of any two numbers is found by finding one number along the top row and the other down the side. Where the column and the row meet you find the product of those two numbers. You can do this backwards, too. Choose a number, say 36. You can see that the numbers which multiply to 36 are 9 and 4. But there are others also: 6 and 6 (and of course 12 and 3). These numbers are called factors of 36. The square numbers are those found when we multiply a number by itself: 1 x 1, 2 x 2, 3 x 3, etc. These form a diagonal line across the chart from the top left-hand corner down to the bottom right-hand corner. Look at the two halves of the pattern either side of this diagonal. They are symmetrical.

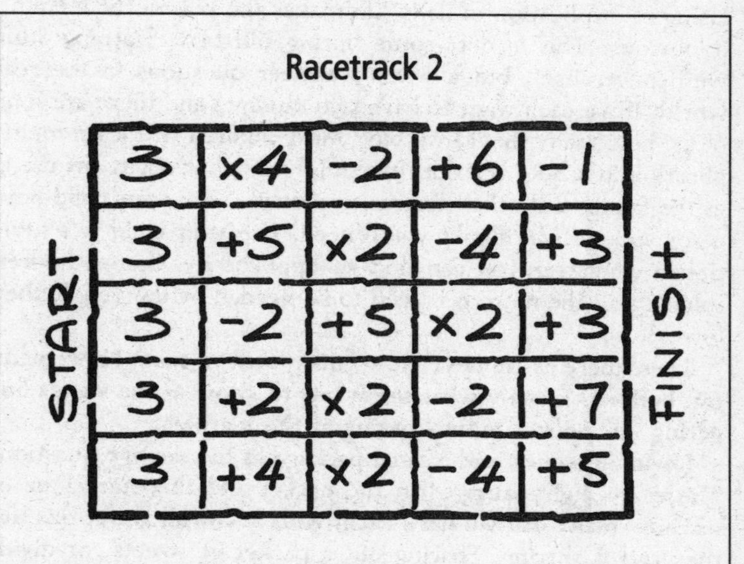


*Racetrack 2: The first to complete the sum is the winner. Compare the answers. Can you find other ways of getting this answer (15)?*

Draw a few straight racetracks alongside each other with a ruler, ending in a finishing line. Mark off squares of equal size all along the tracks and draw a starting line.

This is an 'upgraded' version of the game in Chapter 3, with extra instructions on the squares which read, 'Go back to the start/go back two squares/go forward two squares', etc.

Invite children to choose a little counter or toy to race. Place starters ready. Each player throws the dice in turn and moves the number of squares shown on the upward face of the dice. Who will cross the finishing line first?

Using multiplication in daily life makes the reason for learning it obvious. This is not 'some boring old task, learning stuff we'll never use', but a way to answer questions in the real world. If we each want to have two sausages and there are four of us, how many should we buy? Most children will learn multiplication in school. Parents can simply show how vital its use is in the family and offer chances to practise. Ask your child how many sausages she thinks you'll need. Use your child as a multiplier whenever you can find an opportunity. Some children value a skill the more it is seen to be needed by the people they love.

Then there is money. How much will it cost? How many can I afford? Exactly what she wants to know as she stands fingering her pocket money gazing at the marbles.

Dividing comes in the same package as the sausage question. There are eight sausages in the packet and there are four of us, how many can we have each? Kids know all about this but they call it sharing. Sharing out a packet of sweets, or dividing up a pizza into equal parts seems very important to them. Once again it is the physical doing of a task that brings understanding, with abstraction following in its wake. Once the multiplication tables are known, dividing is automatic because a child can either refer to the chart or will know the factors by heart.

Your child's teacher will probably be the first to teach her to write down 8 divided by 4 equals 2, but check that she knows the division symbol:

$$8 \div 4 = 2$$

One way of looking at dividing is the sharing out in equal parts of all the original quantity. I have 12 sweets and 3 kids. They are all to get exactly the same amount. So the 12 must be fairly shared between them. They each get 4, because I divided 12 by 3. This is the type of example you should begin with because it is so easily understood and demonstrated. You can physically share out the sweets into piles.

However, sometimes division is used in another way. Say you have a large number of carpet tiles. You need to find out how many sets of four you have in the pile of 40 tiles because four cover the width of the passage exactly. You divide 40 by 4 and the answer is 10. Here you are not sharing out the original set between a number of people, you are finding out how many times you can take a smaller set (of four) out of the total (40).

Give your child examples to begin with that do not have remainders or fractions. A strip of paper can be given and the child asked to cut it into equal lengths. Choose a length easily divisible by the number you will ask for—e.g., a strip of paper 21 cm in length will easily divide into thirds, whereas a strip 22 cm in length will present problems to a child asked to separate it into thirds.

Your child can check her answer by going back to the tables chart.

If 12 has to be divided by 4, find 4 down the left-hand column and read along this row until you come to 12. Now look up to the top along this column—the answer is 3.

Another way of checking the answer is of course to multiply 3 by 4. Children may not at first realize that multiplication and division are two sides of the same coin. They may be fascinated to see the neatness with which this 'numbers stuff' all fits together. For some this proves remarkably satisfying.

## NEXT STEPS—REAL LIFE

When your child is confident about multiplying and dividing, explore these ideas found in a sliced loaf of bread

### Maths in a Sliced Loaf

Start by writing down the price paid for the whole loaf. If your child counts the slices, can she work out how much each slice costs?

How heavy is each slice? Weigh one or two to get an average. What is the total weight on the label? Multiply the weight of one slice by the total number of slices. Does this come to the given weight on the label? How many sandwiches of two slices each could be made from this loaf? If you made the whole loaf into sandwiches and each person were to eat two, how many people could you feed? What would be the cost of the bread for each person?

How many slices of bread does your child normally eat per day? So how much do these slices cost, and how much do they weigh?

## Superbuys

Large and multiple packs are supposed to work out cheaper than buying individual packs in smaller sizes. Is this true?

How can you check? You will need to know the price of a smaller pack and its weight, or, for example, the number of pieces of fruit. One kiwi fruit is 17p, but a pack of 8 is 89p. How much does each kiwi fruit in the pack of 8 cost?

An economy funpack of mini chocolate bars costs £1.65 and contains 11 packs. It weighs 110 g. One bar of chocolate costs 25p and weighs 10 g. Which should you buy?

Thinking up questions along these lines is easy around the home. Think about washing powder or dishwasher detergent. If you use 20 g per wash, and the bottle contains 1 kg, how many washes will you get from one bottle?

# 6

. . . . .

# Measuring

Measurement involves length, perimeter and area, scale, capacity and volume, weight, time, direction, and angles. Along with this comes an increased awareness of pattern and shape, comparison and recording. These are a complex group of subjects helped by early play experiences that you can provide at home.

## LENGTH

Have fun exploring length with your child in a number of challenging ways. The games on pages that follow can guide you in the kinds of play that will help your child understand and learn to use the concept of length for himself.

### As Long as...or Is it the Same? (Ages 5 to 8)

Cut two strips of paper the same length. Arrange them parallel.

Ask your child if he thinks they are the same length. Most children say they are. Many children look at the ends of the pieces and if they are in line with one another assume they are the same length. Try moving one of the pieces down a little and asking your child again.

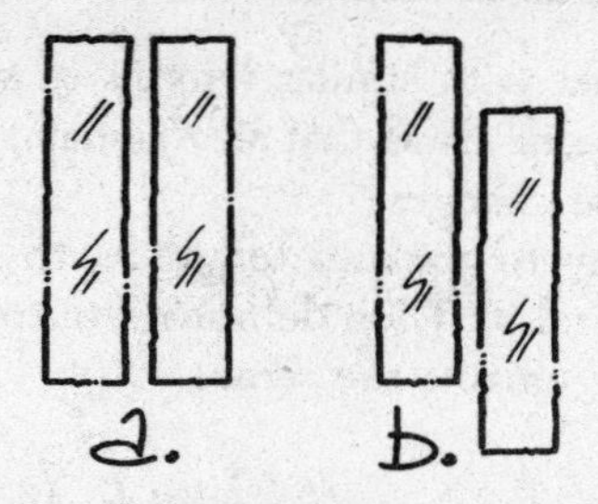


*a. two strips of paper of the same length and parallel, on the same level. b. two strips of the same length and parallel, one moved down.*

Although your child has seen that you used the same two pieces of paper, he may reply that they are no longer the same.

Spend some time together examining these two pieces of paper, measuring them and re-arranging them on the table.

Now fold one of them into a fluted zigzag. Stand it on the table.

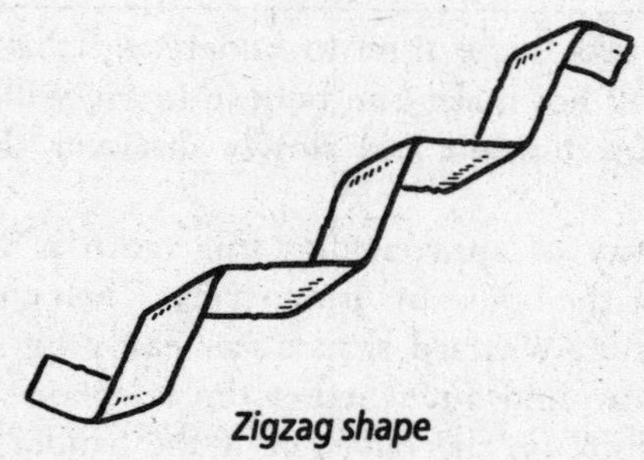

*Zigzag shape*

Are the two pieces of paper still the same length? Give your child the time and opportunities to test out this proof.

Games played with similar lengths of string or ribbon will prove to your child that size remains the same even when the shape changes.

Another way to explain length is to loop a ribbon, twist it and spiral it. Then demonstrate to your child that its length still remains the same.

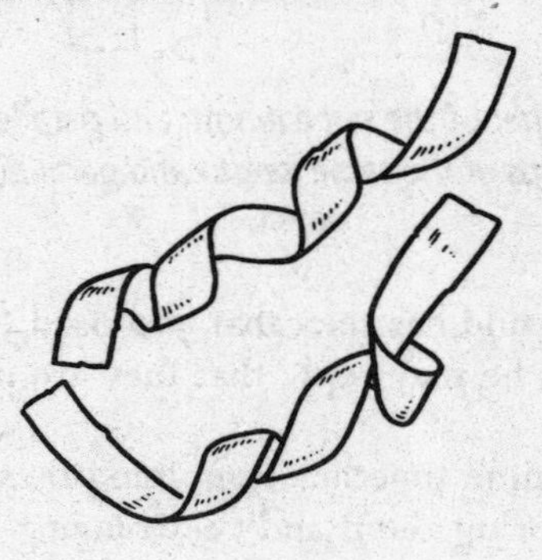

***Paper twisted into a spiral or ribbon***

A child may take some time to understand this fully. Simply telling him will not make him believe it. He will need time to experiment for himself and slowly discover the fascinating truth.

Another way of approaching this truth is by measuring things all over the house by using string. Then check the string against a ruler. Awkward shapes can easily be measured this way, so you can find out whether the telephone receiver is as long as the book on the table, or if the handle on the bag is as long as the arm of the chair. At first glance they may not look equal, because one has a curve. But the string will tell you the answer. Because rulers are confusing at first (there are so many little fine lines), make your own strip of card marking only the centimetres in clear blocks. Make it 10 cm long.

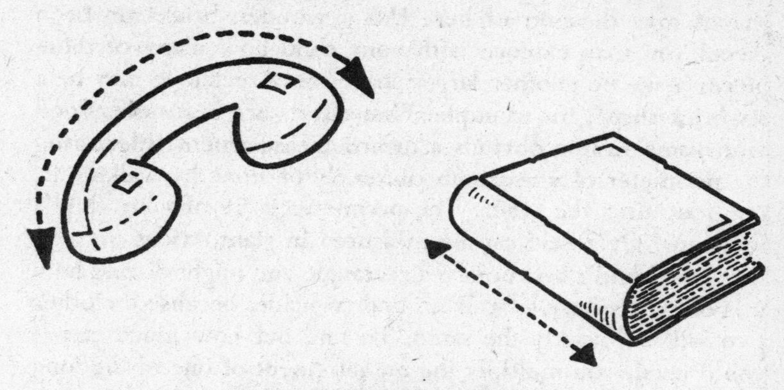

***Measuring things about the house***

A chart could be drawn up to show which things you found were the same, which were the longest, the smallest, and all the stages in between. Was the mug the same height as the pencil?

## PERIMETER AND AREA

The distance all around a plane figure is called its perimeter. So if we want to know how far it is around a playing field, we measure its perimeter, the outer edge all the way round. The space inside this is called the area. This is the grassy area inside the edge of our playing field. We often need to know what these measurements are.

For example, if we're planning a new carpet for Eddie's bedroom, we need to know how much carpet to buy, and how much it will cost. We will need to calculate the area for this. But if we were getting new skirting boards for this room, we'd need to have measured only the perimeter of the room.

Work on these concepts is done when children are fairly

advanced in primary school, but thinking about measuring shapes may develop earlier. Using wooden bricks or Lego pieces you may explore with your child how many of these pieces make up another larger shape. So a rectangle may be a six-brick shape, for example. First efforts are best understood with some simple obvious standard measurement. Measuring the perimeter of a bedroom can easily be done by walking it, one foot after the other. The perimeter is 55 of your child's footsteps. Or a field can be measured in giant strides.

If your child's bedroom is a rectangle you might discuss how it is only necessary to walk around two sides because the other two will be exactly the same. To find out how much carpet you'd need, you multiply the measurement of one of the long sides by the measurement of one of the shorter sides. Length x breadth gives you area.

### Want to Know the Area of Your Foot?

Who could resist this personalized measurement?

Have your child put one bare foot onto a sheet of centimetre squared paper. He now carefully draws around his foot leaving its outline clearly marked on the paper. Now count up the squares together. There will be some half squares—keep a tally of them and add them to the total. Now you know how many square centimetres his foot covers.

## SCALE

Another idea which crops up when thinking about Eddie's carpet is how we might be able to draw a diagram or map of his room on a small piece of paper even though his room is so big. We're going to have to make a certain distance on the paper represent a bigger distance in the room. Let's say we use a centimetre to show every footstep we measured. If we found the length to be 14 footsteps, we'll draw a line on paper of 14 centimetres. At the bottom of our drawing we'll write SCALE: 1 centimetre = 1 footstep.

But there is a problem. Not everyone's feet are the same size and the carpet shop won't know the size of our feet. They'll need a measurement they can recognize. The idea of a standard measurement that everybody understands gives rise to a flurry of ideas about measurement systems. Your child might come up with a few ideas. In the end, though, we have to settle for the one we've got and measure the room in metres. I have a wonderful old metre rule. It's metal, straight and clearly marked and doesn't roll up like an automatic tape which cuts your fingers as it whizzes back into its shell. A length of wood can be marked off in metres, and centimetres. Your child can measure all around the house with this.

Every centimetre on your plan will represent one length of your metre rule. This will not be ultra-accurate, for you'll have to show half metres as half centimetres, and quarters will be tiny. Very small numbers of centimetres will have to be roughly guessed at on the plan. To solve this you write the actual measurement on the drawing. Use squared paper marked in centimetres.

Some people don't mind mixing measurement systems and like to give a child a page of inch squares. Each inch represents the metre rule (or yard if you're old fashioned). These are bigger squares and easier to work in.

Some eight year olds have made me floor plans of their bedrooms, showing the bed and marking in doors and windows. We have made cutouts of furniture to scale and had great fun moving the furniture around in the room and planning different layouts.

## CAPACITY AND VOLUME

Learning about capacity starts with waterplay: endless pouring and filling of containers in the bath and sink, watching the water fill one jug as it runs out of the first. Gradually we find that we can measure this fullness or emptiness. Talk about empty and full. Look at a large bucket and a small teapot. Ask your child to explore how many teapots of water there are in the bucket. Or how much water is three jugs' full. Use a jug that the child can manage easily. Fill it with water and help him pour this water into the bucket. Make a mark to show the level.

Start with concrete experience and allow hours of exploring games. Use grains of rice, beans, sand or water. One of the earliest difficulties may be that children have not yet come to understand that a quantity is the same, or conserved, even when it changes shape. They may believe that water in a tall narrow jug is more than this same water poured into a shallow wide container.

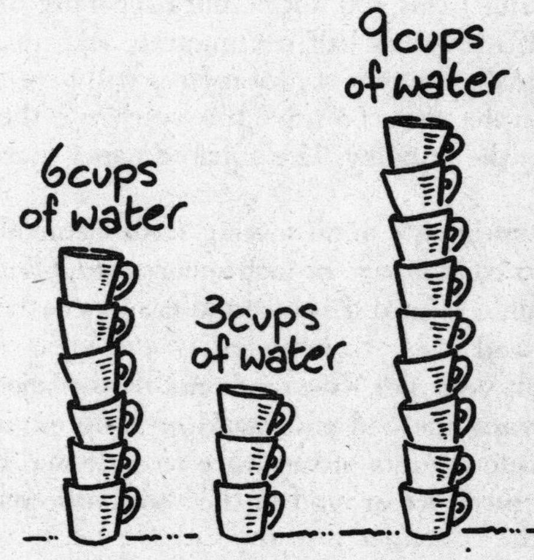


*Bar charts reveal and compare findings*

When we think about a solid figure, the amount of material in it is known as its volume. Choose a cup, mug or jug as your standard measurement. Children can then measure how many mugs a range of big containers will hold. Draw a bar chart to show these findings and compare them. The filling of these containers gives early ideas on volume and capacity.

Comparing leads to ideas such as 'This jug holds 1 cup of water more than the vase,' or 'I can fill all the mugs from this jug.'

When exploring volume it makes things clearer if you play with cubes of a standard size that fit together to make bigger cubes. Use building bricks or other construction sets. Looking at a large cube made up of many smaller ones all the same size is a first step towards introducing what is often a very tough idea for children to grasp.

Once children are confident with the idea of using a fixed standard of measurement, you can introduce the idea of measuring spoons and cups for use in recipes.

## Want to Know the Volume of Your Hand?

Children love experiments that are personal. The fun of being handed a plastic bag squelchy with water in it and being told 'This is your hand' cements this in the memory.

Place a large container filled to the brim with water inside a roasting tin. Check to see that the container is as full as possible and that there is no water spilt into the roasting tin. Now invite your child to plunge his hand into the water up to his wrist. Water will spill over the edge of the container into the tin. He can now take out his hand carefully. The water in the tin is the volume of his hand. Pour this into a watertight polythene bag and seal it. He can play with it. He can pour it into a measuring jug and check how many ml it is.

This is a good way of thinking about the way in which volume remains constant though its shape may change. It also introduces a new way of measuring.

### Displacement

Displacement is always interesting. Fill a measuring jug with water up to a clearly marked level. Drop in various things and note where the water level rises to. When cooking, some people like to measure butter by dropping it into a jug of water. This observation is helpful when you are asked to explain the difference between volume and capacity. Capacity is what an object can hold. Volume is the space it occupies. Looking at a glass of water, you can see that the amount of water it holds is its capacity, while the space occupied by the whole glass itself is its volume. Plunge this glass into your measuring jug as I have described above, and you can measure its volume. Fill the glass with water and then measure this water and you'll have its capacity.

## WEIGHT

Playing around in the kitchen at home is an ideal place to experiment with weight. Try lifting some heavy things like bags of potatoes, or holding objects in your hands, an apple in one and an empty matchbox in the other. Together you can think about how heavy things are, trying out the feel of items by hand at first.

If you have balance scales rather than a fancy modern digital version, you can experiment with getting it to balance. Children will play for hours working on this investigation given a few objects to weigh. It is helpful to include among the objects several identical ones, like bottle tops or marbles, so that your child can work on the idea of equal units adding up to balance either side.

Children always assume that bigger things are heavier. Is this so? How about weighing a cereal box and comparing it to a packet of sugar? Even more startling is an empty cereal box compared to some little pebbles.

## Which Weighs More?

Do potatoes weigh more than cottonreels or plums?

Take one familiar object and have children explore the weight of other items when compared with the first one. Are they heavier or lighter? A potato is heavier than two cherries, it's heavier than a teaspoon, and also heavier than a cottonreel. A teaspoon is lighter than a cup, an apple and a sugar bowl. How many teaspoons would we need to roughly balance the scale with an apple?

Using one item as a constant paves the way for introducing the idea of standard units of weight. It is important to illustrate that the standard measurement of weight (1 kg or 1 lb) never varies but that different amounts of the substances you are weighing are needed to equal these. To begin with you can have your child weigh full packages whose weight is stated on the packaging. This way you can choose weighing tasks that will not produce a fraction of the standard unit. Later you can explain how the smaller units add up to the kg, and measure more accurately.

Your children will love helping with the measuring of ingredients when cooking, because weighing and measuring is a grown-up's game which has a tangible result—a cake or supper. Your patience with flour everywhere and sticky fingers will be rewarded. Either they'll become good cooks, or they'll do well at maths. Perhaps you'll produce a masterchef mathematician.

When a child can weigh fairly accurately, try this experiment:

## Recipe: for Weighing Cookies

You'll need

A balance or kitchen scale that the child can use confidently
2 eggs
Butter
Self-raising flour
Sugar
A little milk
A muffin tin
A mixing bowl

### *Instructions*

- Preheat the oven to 375°F/190°C/Gas Mark 5.
- Have the child grease the muffin tin.
- Now weigh the eggs and make a note of their weight. Set the eggs to one side.
- Weigh out an amount of sugar equal to that of the eggs, then flour, followed by butter. All should weigh the same as the eggs. With some guidance from you at first, children become adept at this recipe and can soon do the weighing for themselves.
- In the bowl, cream the butter and sugar. This is often easier for a young child if you place the bowl on a low table and put a damp cloth or a thin spongewipe beneath the bowl to hold it steady.
- Add the two eggs and beat well. Add the flour and a little milk if needed to make a moist mixture.
- Put spoonfuls into the muffin tin (not filling the hollows more than halfway).
- Bake in the pre-heated oven for 15 to 20 minutes.
- Sift a little icing sugar onto the tops when cool.

## TIME

Small children find the concept of time hard to grasp. At times it seems to fly past and at other times it takes forever for something to happen. The very idea of time marching on at a fixed rate is for some incomprehensible.

You might like to talk about the idea of chunks of time. For example, the week is divided into seven days. Do you do different activities on different days?

- Draw a diagram showing seven days, and some of the things you do on these days.
- You could use a calendar and cross off each day of this week before your child goes to bed.

What did you do yesterday? What will you do tomorrow?

Then come round to shorter periods of time. Let's look at a single day. What do you do first? Talk about the sequence of your day. 'First we wake up and have hugs, then we brush our teeth. Then we have breakfast...'

What time do we go to school/nursery? When do we eat lunch?

Having your child tell about his day in the correct sequence requires him to think it all through in a logical framework.

Draw little pictures of each stage of your child's day and ask him to glue them on to a large sheet in the right order. Talk about what time you do some of these things. If you always get up at seven, draw a clockface and show the time at seven o'clock alongside the picture of waking up. Try to fill in several times that are fixed points in your day. Your child will understand how certain measurements of time order your movements.

There are children who adore alarm clocks or stopwatches and love the idea of measuring how long it takes to eat breakfast, brush teeth and so on. For them this magical grown-up device is an exciting toy from the adult world. When you are talking about hours and minutes, an alarm clock can be fun to play with.

To illustrate how long an hour is you might set the clock to ring in an hour's time. You then all go on with whatever you were doing before and forget about the clock. Suddenly the alarm will go off and you can talk about what you did during the hour, how long or short it seemed and how it is made of shorter time chunks.

Make it clear to your child that one of the chunks of measured time we use is the time it takes for the minute hand to go round the clockface. This is an hour. As each hour passes the short hand moves to the next number on the clockface. But, what about the minute hand?

Now is the moment to cut up a small cake. First cut it in half lengthwise. Each half represents half an hour. Show how the long hand will move halfway around the clock face. Then cut the cake into quarters and demonstrate this too on the clockface. As the time passes a piece of cake (bite-sized) may be eaten every 15 minutes. This image stays in the mind like permapleats in fabric. I have it on record from a very reliable source, now 25, that he still remembers the day I taught him the time this way.

Let your child initially tell the time using only the hours, half-hours and perhaps quarter-hours. Because the cake has no confusing numbers on it, it appears easy to understand the idea of halves and quarters. Much later you can tackle the minutes in between. Your child will notice how the numbers go in order around the clockface. It is often difficult to understand the chunks of time—five minutes each—that are invisibly placed between consecutive numbers which also tell us about hours. Go back to your cake and cut a quarter section into three wedge-shaped slices. Each of these represents five minutes. Repeat the eating game as five minutes pass. Remember the numbers on the clockface tell us about the hour as well as the minutes. Dealing with two ideas at once can be very difficult. Show them separately.

## Beat the Clock

Using an alarm clock or a stopwatch (a kitchen timer will do), play some games against the clock. Set yourselves a time limit and see if you can assemble a Lego model before the timer goes off. Can your child change into pyjamas or dress for school before the timer goes off? Can a group of children keep absolutely silent for one whole minute?

## How Long Is a Minute?

To practice estimating time you might ask your child to tell you when he thinks a minute has passed. You will have set the timer or checked your watch at the start, and if you think he needs help, watch the minute hand go round a clockface for a full minute before beginning. Is it easier to estimate a shorter period of time? What about 10 or 20 seconds? Is there a way he can estimate this? Counting by saying a phrase such as 'Red ones and blue ones' each time, or ' a hundred and one, a hundred and two' will help to get the pace right.

Now ask your child to hold his breath. Press the stopwatch to start it. When he lets his breath out, stop the watch.

Ask him to guess how many seconds passed while he held his breath.

Thinking about longer periods of time, you can talk about years. The family photo album is perfect for this. 'Years ago you looked like this! Years ago Mum and Dad looked like that.' Talk about how many years and put some photos into a sequence. All children love activities based on themselves, so a Story of My Life will be popular. 'This is how you looked four years ago. This is how you look now, four years later.'

Read stories that are based on time sequences. Most local libraries hold a list of children's titles which they recommend for this.

## DIRECTION

Help your child develop a sense of direction. To get to the park, do you have to go North, South, East or West? Look at a map of your area and note North. Discuss the general direction you have to travel to reach the park, or the shops. While on this journey, do you have to turn in other directions?

Draw your own local map or use a large-scale one. Ask your child to help you work out a route you often take together, imagining he needs to explain this to a visitor who is new to the area. Practise this using both 'left' and 'right' instructions and compass points.

### Numberston

A visitor arrives in the town of Numberston (see map on p.76). He gets off the train and asks the stationmaster how to get to the High Street. The stationmaster tells him it is a little far to walk (can you work out how far?), but there is a bus from the station.

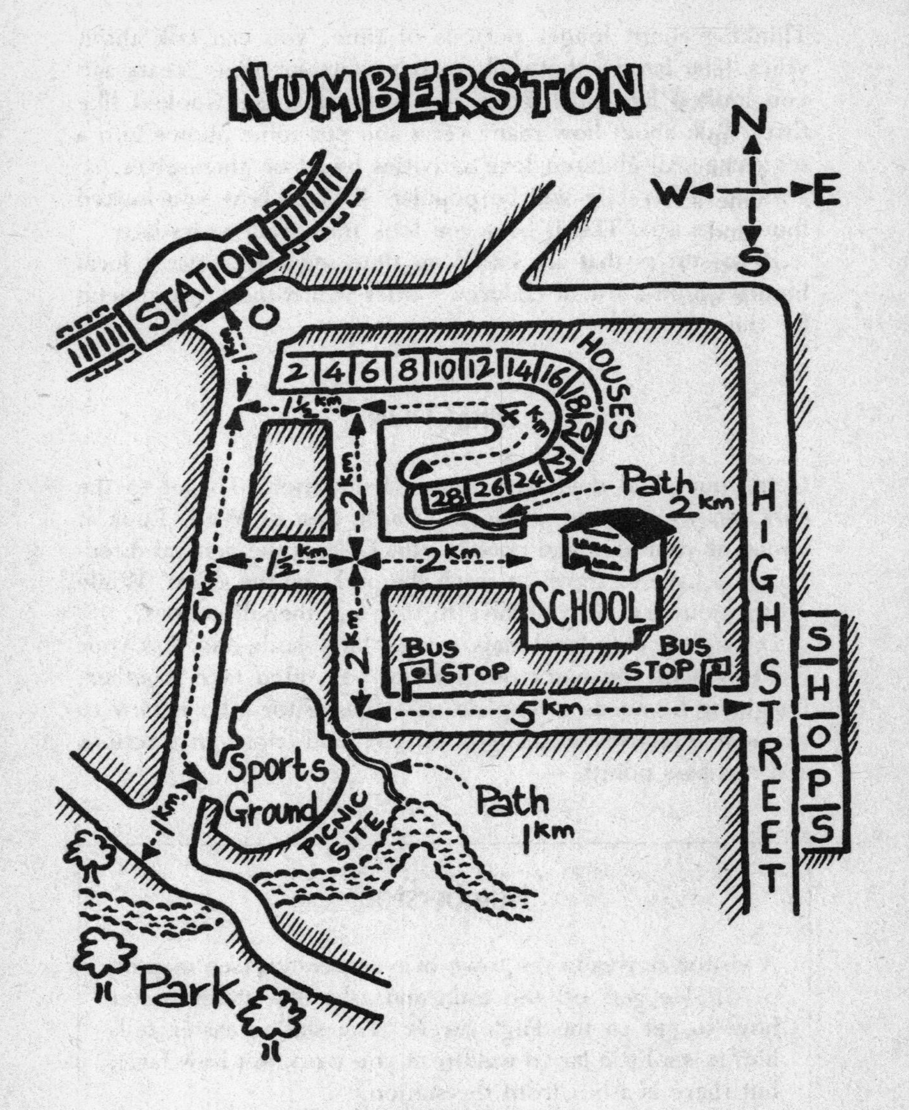


Map of Numberston

Which side of the road should the visitor wait for the bus? What direction will the bus be going in when it picks him up?

Which direction will it be facing when he gets off?

The school building sits in the middle of the school grounds. One side of the grounds is always colder and shadier. The other is sunny. Where would you sit to catch some sun?

## Distance Travelled

Look at the map of Numberston.

In this imaginary town I have hidden many possibilities for questions about numbers. Try out some of these questions with your child or draw your own version.

Can you find out...?

1. how far a shopper must travel from the station to the shops on the High Street?
2. how far a teacher who arrives in Numberston by train must travel to school?
3. Tammy lives in no 8, The Crescent. She could go to school by car. How far would this be?
   Is there another way she could go to school?
   Would this be longer or shorter? By how much?
   Could she go by car on both these routes
   Her mum drives at an average speed of 60 km per hour. What time will they have to leave home to get to school five minutes before the 9 o'clock bell?
4. At weekends Ishmail's family like going to the park. They live at no 2. How far do they travel to the park?

Once in the park, they have a picnic near the bridge by the river. Ishmail's mum decides to leave the family at the picnic and go shopping. She has to walk to the bus stop. How far does she have to walk?

If the bus travels at 50 km per hour, how long will it take her to get to the shops?

The buses come every five minutes. What is the longest her journey might take once she has reached the bus stop?

Use these ideas to create your own investigations. How long is your child's journey to the park, or mum or dad's journey to work?

## ANGLES

Ask a few children what an angle is and you'll get, 'The place where two lines meet' or 'The corner of a triangle'. The simple point you need to make clear is that an angle measures *turn*. Walk along the wall of a room to the corner and then turn when you are forced to at the corner. An angle is the measurement of this turn. Now go to the middle of the floor and turn 360 degrees. Ask your child if he can see that this turn is different from the first one. Give three examples and their correct names. A complete turn is 360 degrees and a half turn is 180 degrees. The first one you did in the corner is (if, as I hope, the walls are at right angles) a right angle or a quarter turn.

## 'Simon Says' Revisited

The old favourite, 'Simon Says', can do a good job of reinforcing angles as you play a few rounds, calling out amounts of turn for the players to make.

## Measuring Angles

A protractor is a wonderful toy that offers hours of exploration as children practise measuring turns. In a busy classroom a teacher may not have the chance to check that each child is using the instrument correctly. Here is where a parent can give some one-to-one attention. Check that the zero line lies on a straight line, with the centre of the protractor at one end of this line. As with all measuring, children need help lining up the zero.

# COORDINATES

## Maps

Look at a few maps. Position are given as coordinates. Set your child searching for a place on the map. Look up the reference in the index and show him how to use the numbers and letters that form a grid on the map. Children enjoy navigating and can learn very early to be dependable navigators.

## Target

A very simple game for younger players.

Draw a three-square grid like the one shown below. Your child has to begin top left and reach the target. He may only travel up, down and sideways. You call out instructions and he moves his counter in the direction you tell him, to the coordinate you give. He moves one square at a time. In this way he reaches his target.

You can play this with a group who play in pairs. Each pair is given a grid. One of the partners in each pair calls out the moves while his partner moves the piece. The pair who reach the target first are the winners.

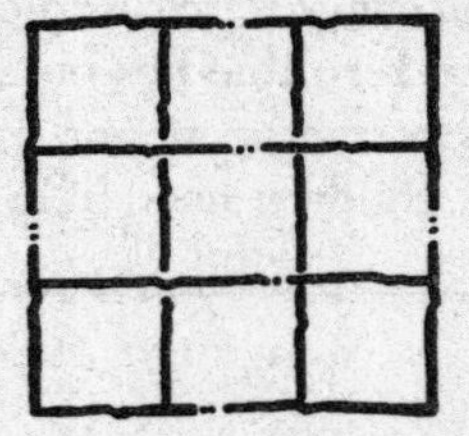

*A three-square grid*

## Robot Man

Here you can make connections with the compass points North, South, East and West, and play games in which one person is given directions by his team to reach a target or treasure, which the team can see on a map you've given out. The player being directed does not see this. They call out instructions like 'Go North three steps, make a quarter turn and go West.' With two teams it's a race to get to the treasure first.

## Treasure Hunt

You are pirates and you have buried your treasure on an island. You decide to leave a map showing the location of the treasure in a bottle. Find a good hiding place. Draw a map on a grid. Write out clear instructions using coordinates. You may also use a secret code if you wish. Simply allocate a number to each letter of the alphabet.

## Drawing by Remote Control

Using two squared paper sheets, number the squares across and give letters to those down the side. Player No 1 draws a picture on her page. She now tries to get her partner to reproduce her picture on his own squared sheet without having seen the drawing. She gives a position on the grid by giving first the reference across, and then the reference down (always this way round in mathematics).

Line by line, positions are given and the player draws a line from one position to the next. Finally the first drawing should appear on the second sheet.

*Coordinates*

## RECORDING MEASUREMENT

Findings and measurements are recorded in different ways. In a scale plan or map we use a small measurement to represent a larger one, perhaps when making a scale drawing of a child's room. But in comparisons we can do some graphic work which helps us see the results at a glance.

Suppose we ask each person in the family which they prefer for breakfast: milk, coffee, orange juice or tea. We can then draw a graphic illustration of what we have found out.

| | MILK | TEA | COFFEE | JUICE |
|---|---|---|---|---|
| Mum | ✓ | ✓ | | |
| Dad | | | ✓ | ✓ |
| Hannah | | | | ✓ |
| Joe | ✓ | | | ✓ |

We can show the drinks by little drawings and make a stroke against each picture to represent someone who chooses this. Ask grandparents, aunts and cousins to build up your numbers. Ask any visitor to the kitchen. Another way of doing this would be to draw little cups, coloured accordingly, and show a cup of juice for every person who chooses this alternative. When you have your finished work, you can count up how many cups of juice you have, and show this as a square in a bar chart.

## Home Projects

Helping with a home DIY project is a chance to learn about angles, edges, areas and measuring accurately. Of course there is also the question of cost. If you're laying a vinyl floor, would the cost of sheeting 3 m wide be very different from tiles at so much a square metre? You pay for each running metre. Would there be some sheeting wasted where you had to make cuts? Would there be less wasted using tiles? If you were to use carpet tiles, how many would you need? A sample tile would be a great help. Keep any tiles left after carpeting or collect a few for maths investigations.

Learning by doing is exciting and rewarding. Simply watching becomes boring.

# 7

• • • • •

# Fractions, Decimals and Percentages

## FRACTIONS AND DECIMALS

Despite fractions being treated as though they were a separate advanced incomprehensible problem by many people, we could choose to see them as an extension of our number system. They simply mean a part of something. When a whole number won't do, we have to talk about parts of a number. We do this in two ways. One way goes by the wonderfully rude name (to children) of vulgar fractions. These are the sort that look like this: ½. The other way uses the decimal system based on ten. The pieces that don't make up a full unit, are fractions. Ten of them will complete a whole number. We always show them as a portion of ten. So a half in this system will be .5 because 5 + 5 = 10; 5 is one half of 10.

The first step, as usual, is to look at your child's concrete experience and what she knows already.

Dividing a chocolate or biscuit in half is an activity kids will be familiar with. You can build up a few observations based on sharing. Cut a fruit into four, cut a round cake or pizza, look at a clockface and remind children how you can tell the half hours. Beginning with what your child already knows seems to make sense. Ask a child how to share something equally and she will do it without hesitation. Now you can give names to these portions. Look at the following diagram and together give names to a half, a quarter, three quarters and then the whole circle. You'll need to give plenty of play and handling opportunities to consolidate this. Explore what happens when you cut a circle into halves, then quarters. How many quarters make up a half? Is your child clear that there are four quarters in a whole?

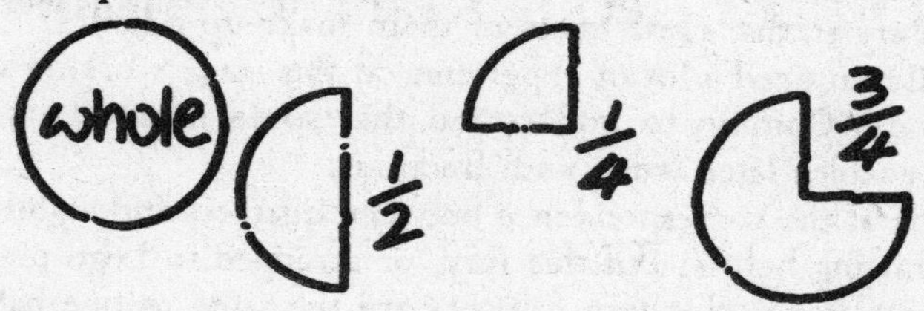


If we cut the quarters in half we have eight pieces, eighths.

There is another way of dividing. Say we have three people and a banana. We can cut it into three equal pieces, which we call thirds.

Fractions turn up when we are dividing and the sum doesn't work out exactly. They are the bits we are left with. But we may want to share them out equally, too, which means dividing them equally into smaller bits.

The top number in a fraction tells us how many things there are to be divided. The bottom number tells us how many shares it will be divided into: ¼.

There are several ways in which we can show a quarter.

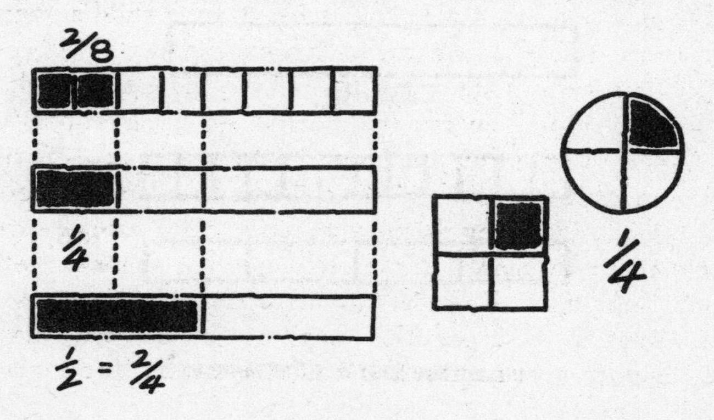

By adding fractions together we can make up a whole. Two halves are the most obvious, but if we have three quarters, what do we need to make a whole? Let your child explore this with wooden bricks, drawings, and cakes or pizzas. Look at the six portions in a processed cheese pack, for example. These are sixths. How many of them make up a half?

Children need a lot of experience at this stage working with fractions. Coming to understand that some mean the same thing enables later work with fractions.

Look at the correspondence between quarters and eighths in the drawing below. Put this way, or arranged in Lego pieces, it is easy to see that two quarters are the same as one half.

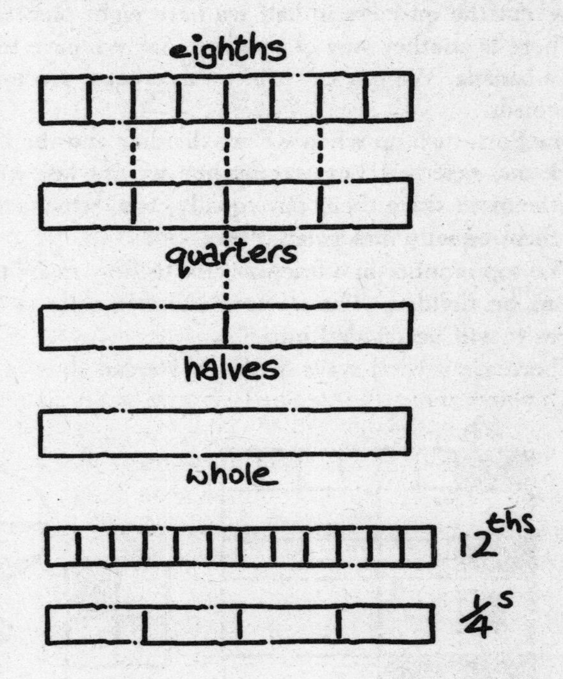


*You can see that ¼ is the same as 3/12*

A chart can be drawn dividing a line into eight equal sections. Above it draw another line exactly the same length and divide it into quarters. Again draw a line the same length above the other two and divide it into halves. Now draw the final line above the others—this represents a whole—1. You can ponder the findings together as you see how $\frac{2}{8}$ are the same as $\frac{1}{4}$ and all the other correspondences.

Demonstrate this with money. 50p is $\frac{1}{2}$ of a £1, 5p is $\frac{1}{2}$ of 10p and 10p = $\frac{1}{10}$ of a £1. So 5p is $\frac{1}{2}$ of $\frac{1}{10}$, or $\frac{1}{20}$ of a £1. Therefore $\frac{2}{20}$ must be the same as $\frac{1}{10}$. When at a later stage your child is doing calculations with fractions, she will have to work with these equivalent fractions. This early work builds a foundation of understanding.

## PERCENTAGES

Fractions, decimals and percentages are different ways of saying the same thing. $\frac{3}{4}$ = 0.75. This is because $\frac{3}{4}$ means three divided by four. Doing this division gives us the answer 0.75.

We can translate this back again, too. 0.75 means $\frac{75}{100}$. When we reduce this fraction to the simplest equivalent fraction, we see that 25 goes into 75 three times, and into 100 four times, so we are left with the answer $\frac{3}{4}$.

Thinking about the decimal system, we might think about how we would invent a new number system. Using the number 10 was the obvious thing to do for a species with 10 fingers and 10 toes. People count on their fingers and the whole base ten counting system seems natural. If you were a Martian with four fingers on each hand, would you possibly invent a different number system?

*Deci* means 10 in Latin, which is where the name 'decimal system' comes from. Put this way, the whole thing seems more friendly and accessible.

Remind your child through some of the games on place

value and base ten (Chapter 4, pages 46–49) how this system works. If we write down a three-digit number—say 453, meaning four hundreds, five tens and three ones or units—we know the value of each digit from its position. Now, what would happen if we were to extend this idea and create places to the right of these three?

453278—we'd need some way of marking a midpoint so that we'd know where the units are. Then we could work out the value of each digit. This is always done with the decimal point. So we could have hundreds, tens, units, tenths, hundredths and thousandths: 453.278. The decimal point, our guide, is always placed after the units column. In a number such as 444.444, can you tell what each 4 represents? When we work with decimals, we always keep this arrangement telling us about place value. To the right of the point are the parts of a whole—the fractions.

Listen to the scores in sports commentaries (on gymnastics or figure skating, for example) to see this system in action. One competitor's score is 9.6; the optimum or best score you can get is 10. This entrant got a score of 9 plus $\frac{6}{10}$, or $\frac{4}{10}$ short of perfection.

Running speeds are often given this way, too. Of course there is something else we handle every day that means understanding decimal places—money. All prices are given in this system—that's how we know whether we are being asked for pounds or pence. £1.50 is recognized by all of us as neither 15 pounds nor 15 pence, but as 1 pound 50 pence.

Can you write the decimal equivalents of:

$\frac{1}{2}$, $\frac{1}{4}$, $\frac{3}{4}$, $\frac{1}{10}$, $\frac{1}{5}$, $\frac{2}{5}$, $\frac{3}{10}$, $\frac{9}{10}$, $\frac{1}{20}$, $\frac{1}{8}$, $\frac{1}{3}$? (Answers at the end of this chapter.)

Some teachers suggest that difficulty with percentages comes when they are seen as an entirely different topic. These teachers suggest too, that if a child has a basic understanding that percentages are simply another form of fractions, she will find later work on them easier.

You might draw a line, and divide it into hundredths. Then divide a second, parallel line into halves and quarters. Your child can see from this the equivalent fractions and 'translate' or decode from ¼ to 25 per cent (or .25). This can be practised as a secret code game.

All percentages are, then, is hundredth parts of a whole. 100 = 1.

10 per cent means ten hundredths of a whole.

$$\frac{1}{2} = \frac{50}{100} = 50\% = .50$$

$$\frac{3}{4} = \frac{75}{100} = 75\% = .75$$

$$1 = \frac{100}{100} = 100\% = 1.00$$

I will not go deeper into fractions or decimals, as some schools do this at different times to others. We are, after all, only the parents, not the teachers. To help a child towards a basic understanding is a positive step, but as there are various ways of tackling the four rules (addition, subtraction, multiplication and division) we should be guided by our children's teachers. If you want to help your child with this, find out when the appropriate time is, and what methods are being taught. Some schools even leave the teaching of fractions until the end of primary school.

*Answers*: 0.5, 0.25, 0.75, 0.1, 0.2, 0.4, 0.3, 0.9, 0.05, 0.125, 0.3333 recurring

# 8

• • • • •

# More Games, Puzzles and Challenges

## BE OBSERVANT

### In the Kitchen

Look to see how many surfaces in the kitchen are horizontal or vertical.

### On Paper

Draw some diagrams.

If two straight lines meet at right angles they are perpendicular, but if they are the same distance apart, alongside each other, they are parallel.

### See this in 3-D

Look with your child at the diagram of the net of a cube on page 91. Could you make your own cube by cutting out this shape in card? Examine the faces of the cube. The idea of parallel can be explored. Look for examples of other ideas you have been examining. Are there right angles? Are the edges perpendicular? Unfold the cardboard box that toothpaste comes in; open it out and see how it folds into its original shape. Look at all packaging afresh.

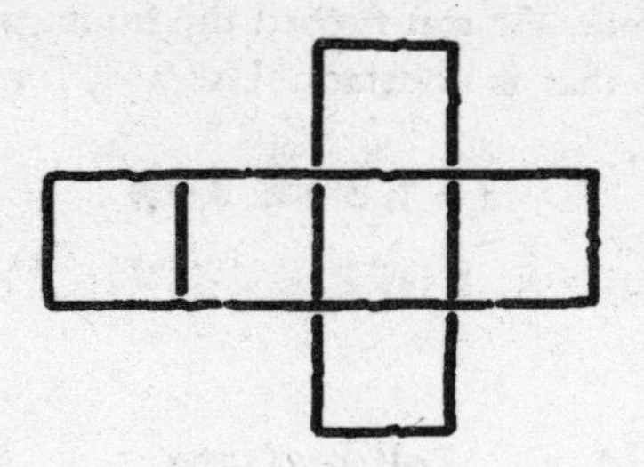

## ADDING AND SUBTRACTING GAMES (AGES 5 AND UP)

### Calculators

Children master calculators as if born with them at their fingertips. They love adding numbers on them and can be set various investigations: Add up the ages of everyone in the family, or add up the numbers in our telephone number.

### Variations on a Theme

You might set them to investigate how many numbers they can make using only numbers 1, 2 and 5. They can either add or take away:

$$5 + 1 = 6,\ 5 + 2 = 7,\ 2 + 1 = 3,\ 5 + 2 + 1 = 8,\ 5 - 1 = 4,$$

$$5 - 2 = 3,\ 5 - 2 - 1 = 2.$$

Using a set of numbered cards, ask your child to find out the totals if you keep one card on the table and add it to each of the other cards in turn, writing down the total of each pair. If 1 is the first card on the table, you then have your child add

2 to it, then 3, etc. He can record the findings on paper. Then change the card that is constant. Use a 3, for example:

**3 + 1, 3 + 2, 3 + 4**

and so on.

## Sell-by Date

In the supermarket, all goods are marked with a sell-by date. Ask your child to advise you. If two yoghurts are dated 16th February and 18th February and today is the 12th, which one should you buy? Why? If you were going to eat it today would there be any reason to take one and not the other? If you were planning on buying a pack of four would your choice be different? If there is one with today's date and it is reduced would it be worth buying?

## Equals

When doing sums your child will soon meet the = sign. This can be easily understood with a balance, the sort of old-fashioned scale with two bowls. Using identical objects, marbles of equal weight for example, or brass buttons, let your child experiment.

Put five buttons into the right-hand bowl. Put three into the left-hand bowl. How many should he add to get the balance right? When he has added two to the left-hand bowl the balance is even, or *equal*. Three and two are the same as five. Three added to two *equals* five.

If you take one away from one of the bowls, you'll have to take one away from the other side too or the balance will be lost.

## Instead of a Crossword—Try Crossnumber!

Draw a diagram like the one shown below. Write the numbers 1 2 3 4 across the top of the page. Invite your child to write these numbers into the squares so that the numbers across add up to five and the numbers down also add up to five. If he solves this in no time, try him out with 5, 6, 7, 8 (which should add to 13). 3, 5, 7, 9 add to 12, while 10, 12, 14, 16 add to 26. Ask if he notices any pattern to his answers. Gradually he will spot that he is always adding the highest number to the lowest, and then the two in between.

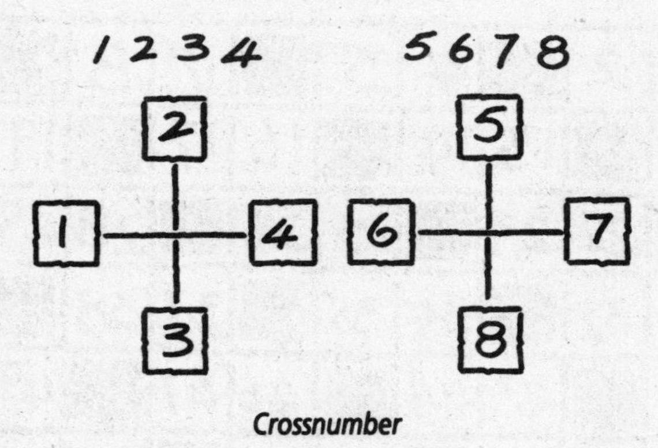


*Crossnumber*

## Brainteaser

A young woman went out shopping and bought a scarf in a market for £1. She paid and left. The next day she returned and said she wanted to exchange it for another. She picked out one worth £2 and started to walk away. The seller asked her quickly for the other pound. But she argued that she had paid him a pound the day before and had returned the scarf worth a pound. Therefore she said, she owed him nothing. Who was right?

## Hidden Pattern

Draw a 10 x 10 grid.

Ask your child to write in the numbers one to ten across the top line. Then colour in any number which is a multiple of three.

Continue on each line, 11 to 20, 21 to 30, etc. until you end with 100 in the lower right-hand corner.

If you colour in all the numbers that are multiples of three, a regular striped pattern will emerge.

Now that you have this square you might like to look at its patterns in various ways. If you coloured in even numbers red and odd numbers blue you'd get a vertically striped square. It is pleasing to look at all those sevens neatly under the number seven, or the fours beneath number four. It puts this counting system into a satisfying coherent diagram.

## SEQUENCE

Number sequences are another form of pattern. These are like hidden secret messages your child will enjoy decoding.

Look at 2, 5, 8, 11—can you tell what comes next? The interval between each number is three.

Here is another sequence which is slightly more complex: 4, 3, 5, 4, 6, 5, 7, This is created by –1, +2, –1, +2. You can set each other sequences to decode, a sequence challenge match against the clock, for example.

## PATTERN PUZZLES (AGES 6 AND UP)

### Triangular Numbers

There is a series of numbers known as triangular numbers. Counters or coins representing these may be arranged in triangular patterns. Take a look at the diagram on page 96.
Triangular numbers

You will see arrangements of the numbers 1, 3, 6, 10 and 15.

Can your child work out what the next number might be? How can this be found? Some people take the numbers themselves—1, 3, 6, 10 and 15—and notice the pattern: they increase by 2, 3, 4 and 5. A sequence.

Other people study the patterns. They simply recognize

them and notice how many extra coins there are each time—2, 3, 4, 5.

Could you predict any triangular number, say the twentieth?

There is another way of looking at this pattern:

The fifth number, which is 15, is the sum of 5 plus the fourth number, which is 10. 10 + 5 = 15.

Fifteen is also equal to 5 + 4 + the third number, which is 6.

It is also equal to 5 + 4 + 3 + the 2nd number, which is 3.

It is also equal to 5 + 4 + 3 + 2 + the first number, which is 1.

So we know that the fifth number is the sum of 5 + 4 + 3 + 2 + 1.

So the twentieth number is the sum of all the numbers from 1 to 20.

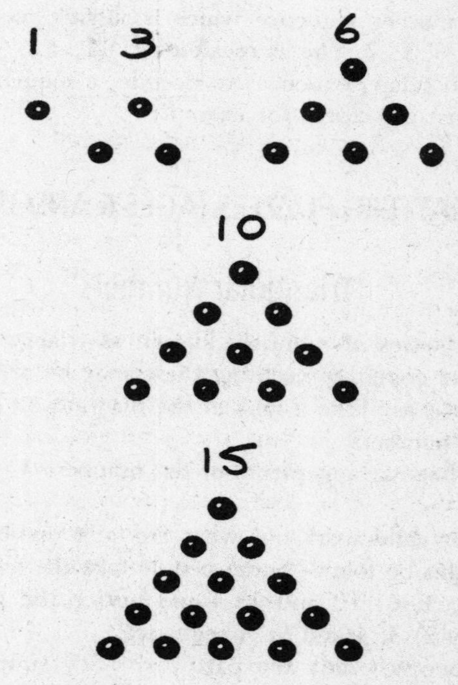

## SYMMETRY

Look at the diagram below. It shows two types of symmetry: reflected symmetry and rotation. You might explore these two by cutting out two identical shapes and playing around with them a little. Try reflecting a pattern in a mirror.

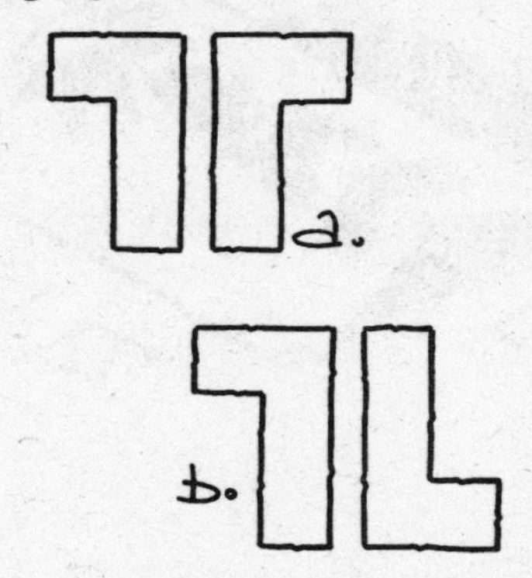


***Symmetry: a. reflected; b. rotated***

Another experiment is to draw a repeat pattern on card and fold it in half exactly on its axis. Place this card on the table and put a small mirror on the table neatly fitting the right angle of the card. The pattern now seems to go on repeating into the distance. Try placing two mirrors at right angles and put the pattern between them: another endless pattern unfolds.

To play around with rotating symmetry you can use any two identical shapes. Make two card triangles or use pieces of processed cheese, segments of an orange, pieces of fuzzyfelt or gummed paper shapes. Make patterns by rotating the pieces.

When you fold a piece of paper twice, with the second fold at right angles to the first, do you have another axis?

## Black and White

*a. Fold a sheet of paper. Cut a series of snips into the fold. Cut away triangles, thin strips, etc. Open out and lay the paper on a darker sheet. b. Fold another sheet and cut your pattern from the edges of the paper. Also try cutting patterns from paper folded in four.*

## ORIGAMI

Learning while you play, the idea at the heart of these games, is not always done through games that are obviously 'maths' to parents. With Origami, the Japanese art of folding paper, a child experiences at first hand what happens when you fold a square sheet of paper accurately in half. Half the square is a triangle; when you fold its tip down you have a trapezoid. These observations come gradually by doing.

For all Origami shapes the starting point is a square sheet of fairly stiff paper. Art shops supply special paper, but fairly substantial typing or copy paper will do. The square must be accurately cut.

All folds must be carefully checked against the edges—sloppy folding does not work.

Ask your child how you might accurately make a square sheet from an A4 rectangular sheet of typing paper.

Fold the top right-hand corner down across to the left edge, creating two overlapping triangles with an uncovered strip below. Cut off this strip.

There are amazing well-known Origami shapes. You may look for these in books on the subject. As a starter, try this graceful bird (continued on next page).

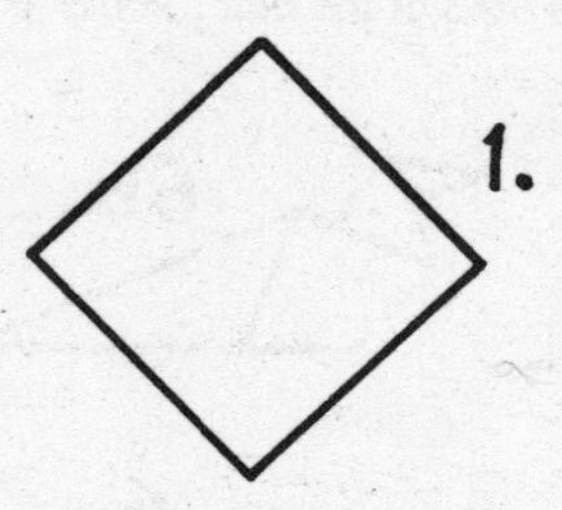


*1. Turn a square sheet of paper so it is in this diamond position.*

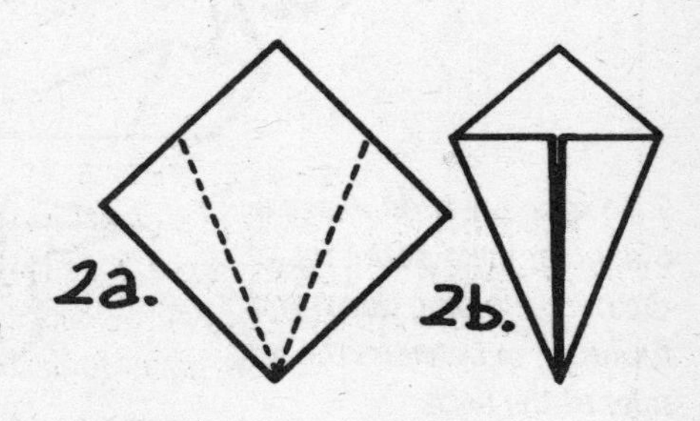


*2. Make triangular folds on each side—your paper should look like a kite.*

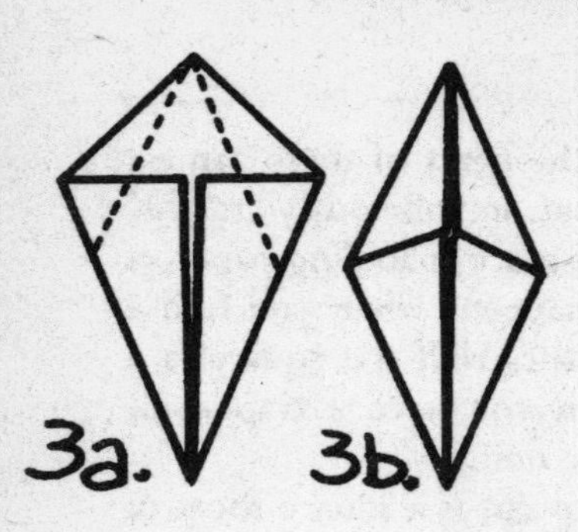


*3. Make two more triangular folds. It should now again look diamond-shaped.*

*4. Turn your diamond clockwise and then onto its side; fold the lower half back.*

5a.

*5. What you've done so far will form the body of your bird. Now it needs a neck. Form a triangular bend at one end of the paper. Fold this triangle upwards. The bottom of this neck should turn inside out as it tucks up between the two sides of the bird's body.*

5b.

6b.

*6. To form the head, make a triangular bend at the top of the neck—folding down and tucking it in between the two sides of the neck.*

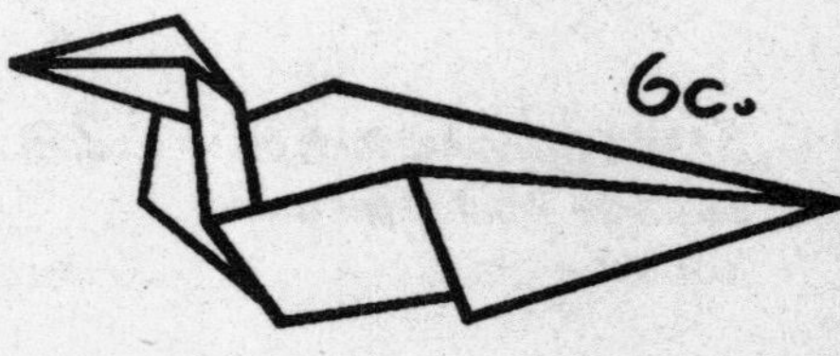

## TANGRAMS

These are ancient Chinese puzzles in which a standard set of seven shapes is rearranged to form recognizable shapes. There is a Tangram cat puzzle and a boat puzzle, a bird and a triangle. There are endless possible shapes.

Cut out shapes accurately from stiff card using the diagram below. Keep the pieces in a large envelope. You and your child may use these pieces any way up.

At first, you might introduce some simple ways in which these shapes build. For example, the two equal triangles (3 and 4) build to a square. Is this the same for square 5? If so, what do we know about the relationship of one of these triangles to the square? It is a half of the square.

Shuffle the pieces and ask your child to recreate the square. Are there other pieces that can build to a square? Numbers 1 and 2, for example? Allow him time to find out.

Using only two pieces, how many shapes can you make? Trace around the shapes you build and see if the other player can make the same shape by copying your traced outline. Gradually move on to three pieces.

Can you use all seven to build a square? Can you use all seven to build a triangle?

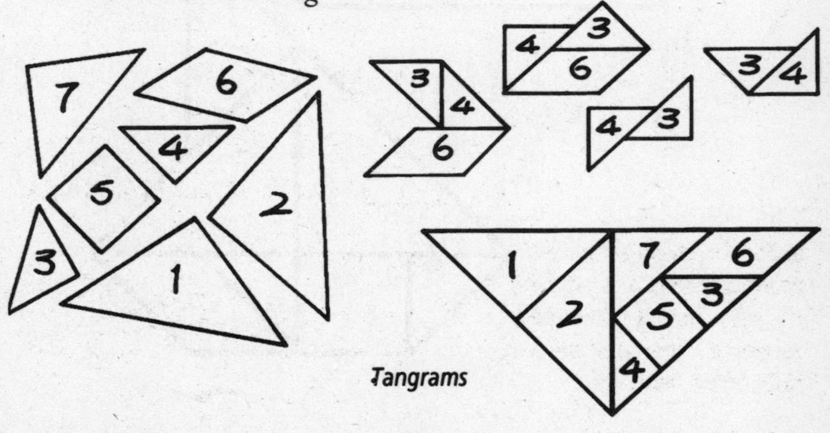


*Tangrams*

7
6
4
5
2
3
1

## TESSELLATIONS

Endlessly fascinating, the patterns created by repeatedly using shapes fitted together to cover a surface are everywhere around us.

Keep a look-out for tiled patterns, repeated geometric patterns, brick paving designs, patchwork and a series of books for children to colour in tessellated patterns created on computer (Altair).

Cut an equilateral triangle from a piece of card. Your child can experiment with this shape by drawing around it or by cutting others from card to work with.

Two triangles form a diamond, six form a hexagon, four can form a parallelogram—there are many possibilities. How could these triangles fit together in a pattern to tile a floor?

Study the design of your imaginary tiled floor carefully. If instead of triangles your tiles were hexagonal, can you see how they would fit together? If you keep a cut supply of cardboard templates your child may find hours of play as they are fitted together in different ways. It is easy to make your own, but patchwork patterns can be bought and floor tiles traced if you need more ideas.

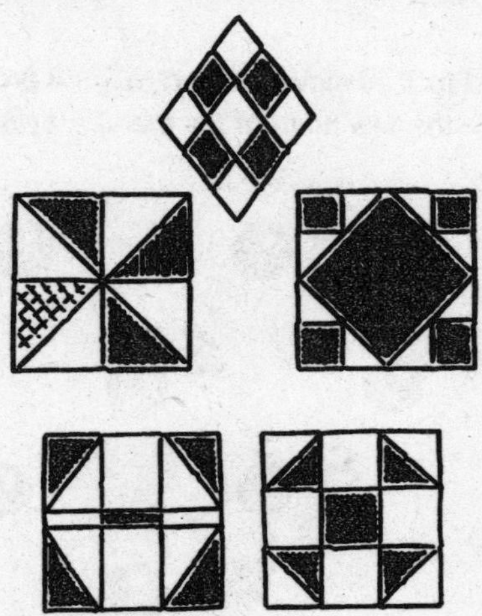

*Patchwork patterns*

## Coin Triangle Puzzle

Try this puzzle with 10 coins.

You may move only three coins and this must reverse the direction of the triangle to make it point downwards.

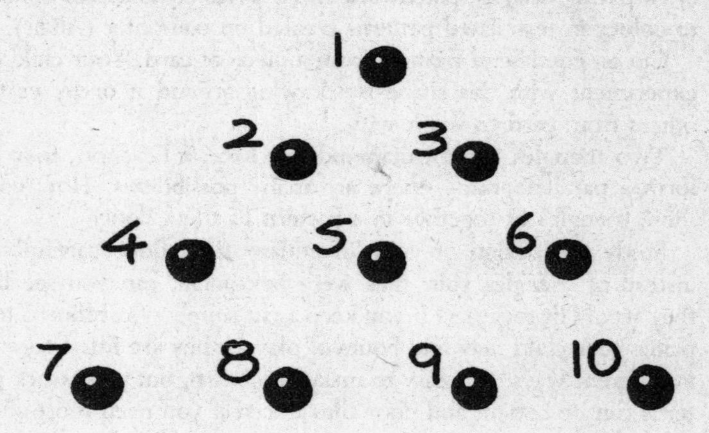


*7 moves up next to 2; 10 moves up next to 3; 1 is brought down to make the new point of the triangle, below 8*

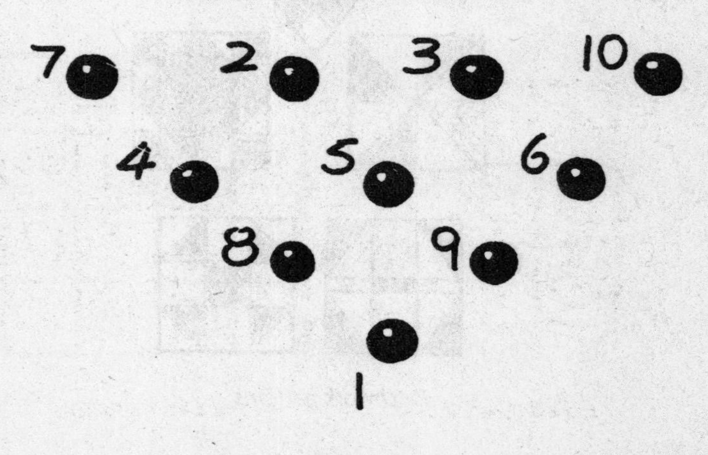

# Magic Squares

A magic square is one of the many fascinating revelations of maths. Said to be discovered by the ancient Chinese from inspiration gained from looking at the markings on a tortoise shell, they were believed to have special magical and spiritual importance. In India, too, magic squares were considered important and a very famous example of a four-by-four square is said to have been known in India about 2000 years ago.

A magic square is a grid of numbers. On this grid, the sum of every row, column and diagonal will produce the same number. In a 3 x 3 grid try and create a layout in which the numbers in each row, column and diagonal adds up to 15. You may use the numbers one to nine.

In a second, larger magic square we have a 4 x 4 grid. You may use the numbers 1 to 16 and make the sum of every row, column and diagonal come to 34.

In the 4 x 4 Indian square shown below right there are other interesting discoveries to make. Try adding up the numbers in each of the four corners. Now look at each little 2 x 2 square which goes to make up the big square. If you add the four numbers in each of these little squares you also get 34.

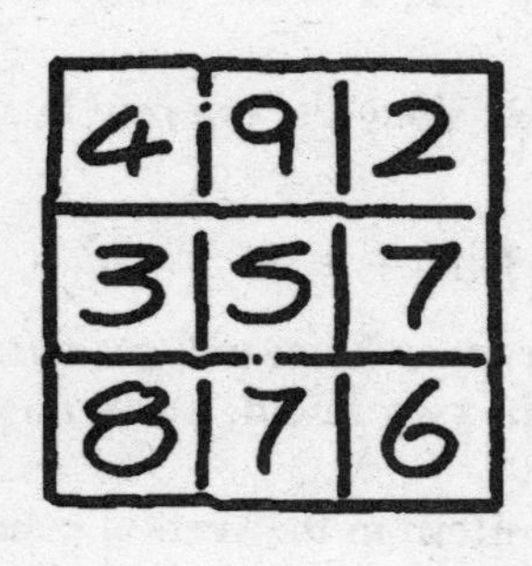


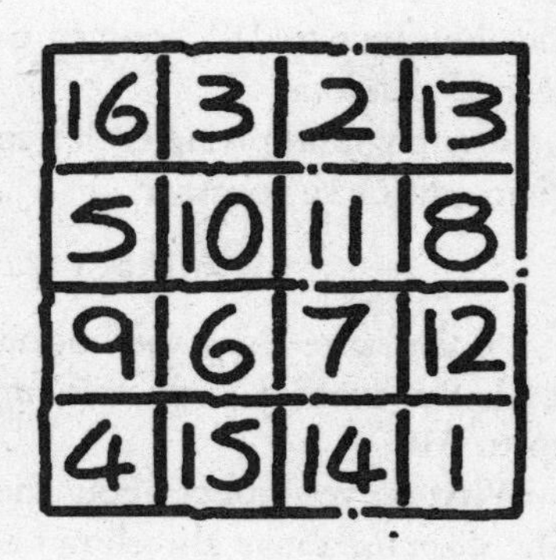

*a. 3 x 3 magic square*

*b. a 4 x 4 square*

*c. a 4 x 4 Indian square*

| 1 | 14 | 15 | 4 |
|---|---|---|---|
| 12 | 7 | 6 | 9 |
| 8 | 11 | 10 | 5 |
| 13 | 2 | 3 | 16 |

| 17 | 24 | 1 | 8 | 15 |
|---|---|---|---|---|
| 23 | 5 | 7 | 14 | 16 |
| 4 | 6 | 13 | 20 | 22 |
| 10 | 12 | 19 | 21 | 3 |
| 11 | 18 | 25 | 2 | 9 |

*d. a 5 x 5 square*

Now try dividing any horizontal row of four figures into two pairs. The top row would be 1 and 14 plus a second pair of 15 and 4. In every row you will find that one pair adds up to 15 and the other to 19. You can easily find a similar pattern in the vertical columns.

Now try multiplying each number along the top row by itself. Then add these together.

$$1 + 196 + 225 + 16 = 438$$

Do the same along the bottom row and you will get the same total. But when you do this for the two middle rows they add up to 310.

What do you notice about the pattern in the vertical columns? The diagram above also shows a 5 x 5 grid. What do each of the columns, rows and diagonals adds up to?

## Combination Lock

See the diagram below. What are the missing numbers?

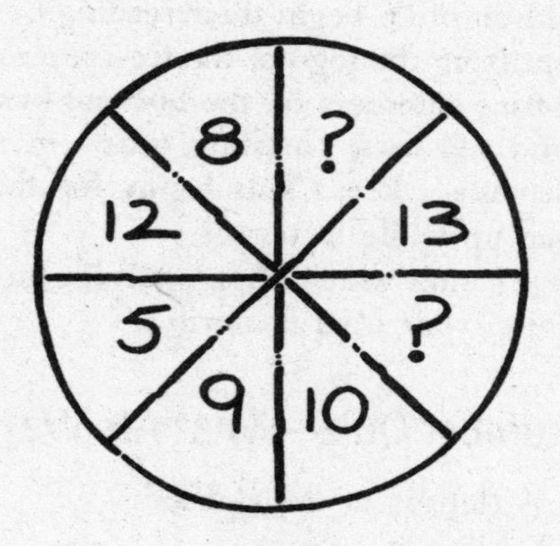


***Answer: opposite segments add up to 18. 9 and 6 are the missing numbers (12 + 6 = 18; 9 + 9 = 18).***

## Paradox from Wonderland—ages 6 and up

A paradox is anything which appears to be true, but is actually false. Or the opposite—it appears false but is actually true. Lewis Carroll has suggested this paradox:

> We can agree can we not, that the better of two clocks is the one which more often shows the correct time? Now suppose that we are offered a choice of two clocks, one of which loses a minute a day, while the other does not run at all. Which one shall we accept?
>
> Common sense tells us to take the one which loses one minute a day, but, if we are to stick to our agreement, we shall have to take the one which doesn't run at all. Why?

## CARMATHS (AGES 6 AND UP)

We spend so much time on the road, in a bus or car, we can play with the numbers, arrows and signs to be interpreted all around us. Children often begin their reading by reading signs on buses or identifying the logo of the ice-cream or hamburger they crave. Spotting numbers on the bus and bus stop, counting blue cars and red cars, counting taxis—in short, playing what I call 'carmaths', keeps kids happy for those miserable moments cooped up while in transit.

Here are a few games which I hope will act merely as a basis for you to develop your own favourites.

### Number Quiz—What Are We?

1. 26 L of the A (letters of the alphabet)
2. 365 D in a Y (days in a year)
3. 100 P in a P (pence in a pound)
4. 60 S in a M (seconds in a minute)
5. 7 D of the W (days of the week)
6. 9 P in the SS (planets in the solar system)
7. 3 BM (blind mice)

I'm sure you get the idea and could produce your own!

### Numberplate Maths

- Can you spot a numberplate with all even/odd numbers?
- Can you spot a numberplate whose numbers add to ten?
- Who can spot the numbers 1 to 10 consecutively first?

### On the Road

Count cars of different colours and compare totals. By how much do they differ?

## Estimate

Played in the car, make sure this does not disturb the driver. Pick a landmark in the distance, a railway bridge, a tower or a spire. The children shut their eyes and shout out 'Now' when they guess you've reached it.

## Navigate

Keep them absorbed with a navigation task. If the Ordnance Survey map is to difficult, draw a simple map for younger children showing a few major easily spotted landmarks—left at the bridge, right at the petrol station, straight on past the railway station and over the river. Our daughter's skill, once developed, has got us out of a maze of streets more than once.

## Compass Reading

A compass is not only fun to play with in a car, it's a vital necessity if you're walking or cycling. Orientating yourself is a good life skill to have—start them young with the adventure of a grown-up task and children will be very willing to join you in working directions out.

## Spot the Road Sign

First you'll need a guide to or list of road signs. Players spot them on the road and look them up. With a pencil they tick each time they get one. The player with the most spotted and explained is the winner. They soon come to know the most common ones and will know what they mean.

## Guessing Games

Any guessing game played in your head, described elsewhere in this book, can of course be used in the car. Finger Guessing Games, Guess the Number, Think of a Number, Add these... and Times Tables are just a few.

## Matches

Waiting around? A few matches in your pocket can provide puzzles. They can be used for counting, bundling in tens, making into squares or triangles. Try this:

Arrange twelve matches in four adjoining squares to create one bigger square. The puzzle is to find how you can take away only two sticks but leave two complete squares (see diagram for answer!).

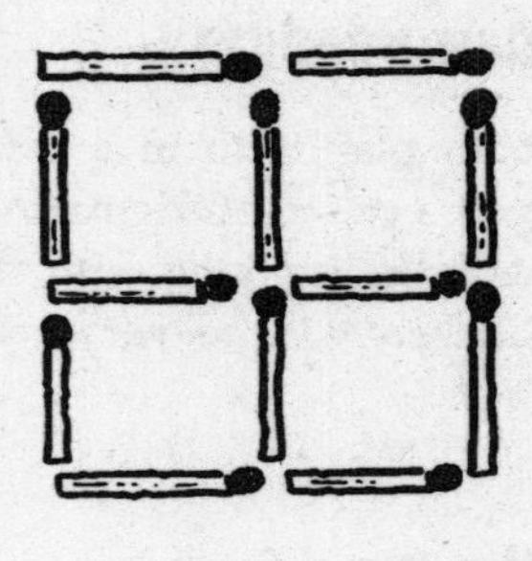

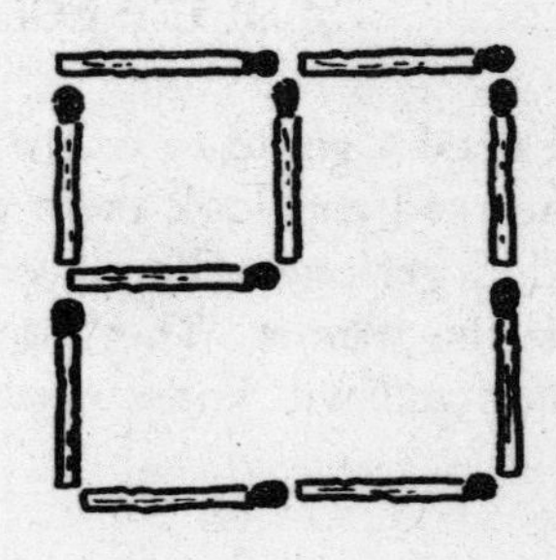

## Pub Legs

This is an old favourite with my family. We play this in the car. Each time a pub is passed the players take a quick covert look at the sign. The object of the game is to accumulate the highest score of legs. You do this by adding up the legs in the sign. For example, 'The Horse and Cart' would yield four legs. When there is a name with many horses, e.g. 'The Coach and Horses', you are allowed to take four horses. So this sign gives 16 legs. 'The Fox & Hounds' would give the fox and his four legs plus four hounds each with four legs. Total 20. Why I said you should look covertly at the sign is because you rather hope your opponents have not spotted it. As soon as you can, call out your total and tell the others the name of the pub. This last step is to prevent players inventing pubs that don't exist!

Five little soldiers,
On a cellar door,
One tumbled in,
And then there were four.

Four little soldiers,
Climbing up a tree,
One broke a leg,
And then there were three.

Three little soldiers,
Out in a canoe,
One tumbled out,
And then there were two.

Two little soldiers,
Fooling with a gun,
One shot the other,
And now there is one.

One little soldier,
With his little wife,
Lived in a castle,
The rest of his life.

Three blind mice,
See how they run,
They all ran after the farmer's wife
She cut off their tails with a carving knife,
Have you ever seen such a sight in your life?
As three blind mice.

Ten green bottles,
Hanging on the wall.
There are ten green bottles,
Hanging on the wall.
And if one green bottle, should accidentally fall,
There'd be nine green bottles, hanging on the wall.
Nine green bottles,
Hanging on the wall...

One, two, three four five,
Once I caught a fish alive,
Six, seven, eight nine ten,
then I threw it back again.

Five currant buns in the baker's shop
Round and fat with sugar on the top.
Along came a boy with a penny one day
He bought a currant bun and took it far away.
Four currant buns in the baker's shop...

Five fat sausages sizzling in the pan,
One went pop and then it went bang.
Four fat sausages sizzling in the pan,
One went pop and then it went bang.
Three fat sausages sizzling in the pan...

# APPENDIX 2

. . . . .

# Hands-on Fun

Playdough is without doubt the cheapest, most satisfying constantly changing play material you can produce. Make up a fresh batch from time to time and store it tightly wrapped in the fridge. Make numbers with it, make creatures with numbers of legs, make cylinders...explore shapes. The relaxing squelchy feel of the dough in little hands is soothing and will often quieten down a tantrum, a fight or general chaos.

## HOMEMADE PLAYDOUGH

1 tablespoon cooking oil
1 cup flour
½ cup salt
1 cup water
2 teaspoons cream of tartar
a few drops of food colouring to colour

Mix to a smooth paste. Cook slowly in a pan for a few minutes. Cool, wrap well and chill.

## FINGERPAINTS

There is a freedom in fingerpainting unequalled by any other method. Draw a number and wipe it away, make a pattern and swirl it into another one. Paper with a shiny texture is ideal to work on but so are many other surfaces. On sunny days we would fingerpaint the outside kitchen steps and then simply hose down when we were done. Use plastic sheets, perspex, baking sheets, plastic trays, formica, fridge doors (if you can stand the mess), etc....

½ mug instant cold water starch
½ mug soap flakes (Lux)
⅝ mug water
powder paint as needed

Beat the mixture until it looks like whipped potato. Add colour (powder paint). Wet the chosen surface with a damp sponge. Spread the paint onto the wet surface. Spread with a spatula, wooden lolly stick or tongue depressor (obtainable from the chemist's). Start to use your fingers. Draw and wipe away (with the sponge) like a magic slate. If you wish to keep a design, make a print of it by laying a sheet of paper over the design and pressing down. A print taken in this way can then be hung up to dry.

## SPRAYPAINTING OR SPATTERING

This technique is fun to try when you are working with number shapes. Cut a number from card or stiff paper. This is your template. Lay it on a sheet of paper and have your child spatter paint over it. When he is finished, lift up the number template and your wonderful number will be the colour of the paper while around it is a whirl of fine droplets, often in several colours.

The opposite of this is to use the card you cut your number from, which will now have a number cut out of it, as a stencil. Lay this stencil on your paper. Spatter away. Only the cut-out number space will be filled with fine drops, the space around it remaining untouched.

## Spattering Tips

Spattering goes everywhere. Here are some tips for maximizing the fun and minimizing the mess:

- Cover the child in an old adult shirt or have him do the spattering outside with no clothes on.

- A large cardboard box is a great help. Lay the work on the bottom of the box and the overspray will be contained in the box and on its sides rather than everywhere else.

- Place powderpaint made up fairly wet into plastic tubs (old margarine tubs will do).

- Have your child use an old toothbrush or nail brush to dip into the paint and then—with his fingers, a lolly stick or ruler—pull back the bristles in a steady stroke aiming towards the paper. Help him at first so that paint does not go into his face. Spatterers become adept very quickly.

- Try mixing a few colours that blend well. Swirls of purple with pink and lilac and blue are harmonious.

## WAX RESIST

Have your child draw a number with a white wax crayon on a sheet of white paper. Now he paints over the sheet with powder paint and a generous brush. The waxed area will not take the paint, so your number will show up.

*Printing* and *Rubbing* are described in the text, but I remind you of them here as other techniques that are both fun and instructive to young mathematicians.

## APPENDIX 3

• • • • •

# Books to Help with Counting

Books form a bridge between adult and child and a bridge to the outer world. They also make a learning experience a true adventure. You will have ABC books, and nursery rhyme collections—Mother Goose volumes with enchanting pictures, children's verse books with humour and wonderful story and picture books. Poring over these together you'll spot many opportunities to count, talk about numbers, tell the story in your own words, getting the sequence of events in the right order. You'll find stories which emphasize the days of the week or the passing of time. Without making heavy going of it, you'll find reinforcing your number explorations easy with books.

For the parent who says his child won't 'settle to learning', read him a story in the bath where he can play and splash while he listens.

Below is a brief list giving only a hint of the wealth and artistry in books that help with some of the ideas in this book.

Janet and Allan Ahlberg, *Each Peach, Pear, Plum* (Puffin Books). We could never be without this, or indeed any of the other classic Ahlberg collaborations, such as *The Baby's Catalogue* and *Starting School*.

Mitsumasa Anno, *Anno's Counting Book* (Bodley Head). Offers times of day, seasons and mathematical relationships in everyday living. Anno has a series of mathematical books including *Anno's Mysterious Multiplying Jar* and *The Earth Is a Sundial*, all published by Bodley Head.

Jeannie Baker, *One Hungry Spider* (Deutsch).
A 3-D collage counting book of immense innovative power.

Stan and Jan Berenstain, *Inside, Outside, Upside Down* (Collins).
A favourite with many children.

Eric Carle, *The Very Hungry Caterpillar* (Hamish Hamilton).
A classic. Within the story and the glorious metamorphosis of the caterpillar, we find the days of the week, counting, and the enchanting peep holes.

Babette Coles, *The Hairy Book* (Bodley Head).

—, *The Smelly Book* (Bodley Head).

—, *The Slimy Book* (Bodley Head).
Coles' imaginatively horrible books invite descriptions as awful as you can manage. Description is, strangely, very important in maths.

R. Crowther, *The Most Amazing Hide & Seek Counting Book* (Kestrel).
Has creatures springing out and popping up.

Susanna Gretz, *Teddybears 1–10* (Black).
A counting book.

Pat Hutchins, *1 Hunter* (Bodley Head and Picture Puffin).
A 1 to 10 counting book. You probably already have a few Pat Hutchins books on your shelf.

MacMillan Education's 'First Skills' series has stories linking everyday events to a specific theme, includes *Opposites*, *Shapes*, *Time* and *Counting*.

Colin McNaughton, *Books of Opposites* (Methuen).
A series of five books.

—, *Seasons* (Methuen).
Another group of four in a boardbook series.

Rodney Peppe, *Odd One Out* (Viking Kestrel).
Very useful.

Jan Pienkowski, *Concept Books* (Picture Puffin).
These include ABC, Colours, Numbers, Shapes, Sizes and Time among others.

Charlotte Pomerantz, with Jese Aruego and Ariane Dewey (illus.), *One Duck, Another Duck* (Julia McRae).
Helps with counting.

Fiona Pragoff's photobooks are artworks not to be missed. These are simple boardbooks with superb photographs, one page illustrating the theme. Plenty to talk about and gaze at here.

John Prater, *On Friday Something Funny Happened* (Bodley Head).
Highlights time sequence and days of the week.

William Stubbs (illus.), *This Little Piggy* (Bodley Head).
The old rhyme brought to life.

C. Watson and D. Higham, *Opposites* (Usborne).
Part of Usborne's 'Simple Facts' collection, which also includes *Shapes* and *Sizes* by the same authors.

Garth Williams and Patrick Hardy, *The Chicken Book* (Picture Lions).
A counting rhyme.

# Index

• • • • •

addition 40–49
  games 91–4
angles 78–9
area 63–4
arranging game 40–1
arrows game 19
as long as .... game 60–62
automatic counting 54

base ten games 46–9
beat the clock game 74
brainteaser 93
build 10 game 45
butterfly blobs 16

calculator games 91
can you find? game 9
capacity 66–9
carmaths 108–11
classification skills 7–8
coin triangle puzzle 104
combination lock game 107
compass reading 109
concentration 3, 25
confidence–building 2–3
contrasting 7–9
cookie recipe for weighing 71
counting 26–9, 54
counting games 33–4
counting rhymes 34, 114–18
crossnumber puzzles 93
cubes as 3-D shapes 90–91

decimals 84–9
dice games 35–6, 44, 48
differences, distinguishing 7–8
direction 75–8
discovery learning 25
displacement 69
distance travelled games 77–8
division 56–8
Doman, Glen 23
dominos 27–8
drawing by remote control game 81

early action strategies 6
equals 92
estimation games 74, 108
experimental learning 3–5

fantasy games 38
fingerpaints 118
flashcards 23–4
foot, area of 64
fractions 84–7

giant's castle game 52
guessing games 110

hand, volume of 67–8
hidden pattern 94–5
home projects 83
how long is a minute? game 74
hundreds, tens and units 46–8

kitchen games 90

language skills for number 16–18

length 60–63

magic beans game 41
magic squares 105–6
maps 79–80, 109
match puzzles 110
matching 7–9
measurement 60–70
money 57
multiplication 50–59
multiplication tables 53–6

navigation games 79, 109
number bonds 43
number circles game 51
number lines 42–3
number quiz 108
numberplate maths 108
Numberston games 75–8
numerals:
  names of 21
  shapes of 31–3
  written symbols 21, 23, 24
nuts at the zoo game 52

observation 3–4
Origami 99–100
over–enthusiasm, from adults 4

painting games 16, 118–20
pairing 7–9
paper drawing games 90
paper and scissors games 98
paradoxes 107
patterns 12–16, 49
  hidden pattern 94–5
  puzzles 95–6
percentages 87–9
perimeter 63–4
personalised rhymes 29–30
place settings game 38
place value games 46–9
playdough 117
position words 18–19
praise 5
prints 15–16
pub legs game 111

quantities, recognising 23, 27
questions, answering 5, 25

racetrack games 35–6, 55–6
real life uses 37–9, 58–9
recording measurement 82–3
relationships 12–14
road sign spotting game 109
robot man game 80
rollerdice game 48
rubbings 15–16

scale 64–5
sell–by date game 92
sequences 95
shapes 10–11
  games 99–103
  of numbers 31–3
shapes by touch game 10
'Simon says' games 20, 79
size games 99–103
sliced loaf game 58
sorting 7–10
sounds 14
spattering 118–19
spraypainting 118–19
stranger game 11
subtraction 40–46
  games 91–2
superbuys game 59
survey game 37
symbols 21
symmetry 12, 97

tables 53–5
Tangrams 101–2

target game 80
teatime game 37
tessellations 103
think of a number game 46
thinking skills 16–20
three bears game 52
time:
  concept of 72–5
  for learning sessions 24
treasure hunt games 39, 81
triangular numbers 95–6

UFOs game 18

volume 66–9

watching 3–4
wax resist 120
weight 69–71
which weighs more? game 70
word collections 17